Springer Theses

Recognizing Outstanding Ph.D. Research

Aims and Scope

The series "Springer Theses" brings together a selection of the very best Ph.D. theses from around the world and across the physical sciences. Nominated and endorsed by two recognized specialists, each published volume has been selected for its scientific excellence and the high impact of its contents for the pertinent field of research. For greater accessibility to non-specialists, the published versions include an extended introduction, as well as a foreword by the student's supervisor explaining the special relevance of the work for the field. As a whole, the series will provide a valuable resource both for newcomers to the research fields described, and for other scientists seeking detailed background information on special questions. Finally, it provides an accredited documentation of the valuable contributions made by today's younger generation of scientists.

Theses are accepted into the series by invited nomination only and must fulfill all of the following criteria

- They must be written in good English.
- The topic should fall within the confines of Chemistry, Physics, Earth Sciences, Engineering and related interdisciplinary fields such as Materials, Nanoscience, Chemical Engineering, Complex Systems and Biophysics.
- The work reported in the thesis must represent a significant scientific advance.
- If the thesis includes previously published material, permission to reproduce this must be gained from the respective copyright holder.
- They must have been examined and passed during the 12 months prior to nomination.
- Each thesis should include a foreword by the supervisor outlining the significance of its content.
- The theses should have a clearly defined structure including an introduction accessible to scientists not expert in that particular field.

More information about this series at http://www.springer.com/series/8790

Junpei Matsuoka

Total Synthesis of Indole Alkaloids

Based on Direct Construction
of Pyrrolocarbazole Scaffolds
via Gold-Catalyzed Cascade Cyclizations

Doctoral Thesis accepted by
Kyoto University, Kyoto, Japan

Springer

Author
Dr. Junpei Matsuoka
Faculty of Pharmaceutical Sciences
Doshisha Women's College of Liberal Arts
Kyotanabe, Japan

Supervisor
Prof. Hiroaki Ohno
Graduate School of Pharmaceutical Sciences
Kyoto University
Kyoto, Japan

ISSN 2190-5053 ISSN 2190-5061 (electronic)
Springer Theses
ISBN 978-981-15-8654-5 ISBN 978-981-15-8652-1 (eBook)
https://doi.org/10.1007/978-981-15-8652-1

This Springer imprint is published by the registered company Springer Nature Singapore Pte Ltd.
The registered company address is: 152 Beach Road, #21-01/04 Gateway East, Singapore 189721,
Singapore

Supervisor's Foreword

It is a great pleasure to introduce Dr. Junpei Matsuoka's thesis work on the Springer Thesis Prize as an outstanding original doctoral work. Dr. Matsuoka graduated from Meijo University in March 2016. In April 2016, he entered the Graduate School of Pharmaceutical Sciences, Kyoto University, and obtained his Ph.D. degree in my group in March 2020.

In recent years, diversity-oriented total synthesis of natural products based on direct construction of the core structures has been recognized as an important subject in organic and medicinal chemistry. The main part of Dr. Matsuoka's doctoral study is diversity-oriented total synthesis of dictyodendrins using a gold-catalyzed cascade reaction for the construction of the core structure, the pyrrolo[2,3-c]carbazole scaffold. After succeeded in the key cascade reaction, he faced difficulty in introduction of the requisite substituents into the resulting electron-rich pyrrolocarbazole scaffold. He suffered from the low reactivity of the Boc-protected pyrrolocarbazole and instability of the corresponding unprotected derivate which readily underwent rearrangement and polymerization. After considerable experimentations, he fully exercised his patience to overcome this difficulty and achieved the diversity-oriented total synthesis of several dictyodendrins based on late-stage functionalization. He published these significant results in *Angewandte Chemie* in 2017 and *Chemistry European Journal* (CEJ) in 2020 as the first author. Notably, it has been confirmed that the CEJ paper has been accepted on the cover page, and the profile of the research team is going to be introduced. Furthermore, Dr. Matsuoka has been working on the development of the direct construction of aspidosperma alkaloids and succeeded in formal total synthesis of vindorosine, which has been published in the *Journal of Organic Chemistry*.

Dr. Matsuoka has outstanding ability to complete natural product synthesis. The difficulties during the total syntheses can be solved by his insight to follow the reaction and logically analyze the experimental results, as well as the mental and physical strength. He was selected for the JSPS research fellowship for Young Scientists and Young Researchers Overseas Challenge Program and won some important awards such as the Kansai Branch Award, the Pharmaceutical Society of Japan (PSJ), for Young Scientists.

His thesis study has shown that gold-catalyzed cascade cyclizations have a significant power for direct construction of the core structures of biologically active alkaloids, which accelerates their total syntheses and medicinal applications. I hope his outstanding thesis will contribute to synthetic research of many readers.

Kyoto, Japan   Prof. Hiroaki Ohno
July 2020

List of Publications

This study was published in the following papers.

Chapter 2.
Total Synthesis of Dictyodendrins by the Gold-Catalyzed Cascade Cyclization of Conjugated Diynes with Pyrroles Junpei Matsuoka, Yuka Matsuda, Yuiki Kawada, Shinya Oishi, Hiroaki Ohno *Angew. Chem. Int. Ed.* **2017**, *56*, 7444−7448.

Total Synthesis of Dictyodendrins A–F by the Gold-Catalyzed Cascade Cyclization of Conjugated Diynes with Pyrroles Junpei Matsuoka, Shinsuke Inuki, Yoichi Miyamoto, Mayumi Otani, Masahiro Oka, Shinya Oishi, and Hiroaki Ohno *Chem. Eur. J. in press.* (https://doi.org/10.1002/chem.202001950).

Chapter 3.
Construction of the Pyrrolo[2,3-*d*]carbazole Core of Spiroindoline Alkaloids by Gold-Catalyzed Cascade Cyclization of Ynamide Junpei Matsuoka, Hiroshi Kumagai, Shinsuke Inuki, Shinya Oishi, Hiroaki Ohno *J. Org. Chem.* **2019**, *84*, 9358–9363

Acknowledgements

The author would like to express his sincere and wholehearted appreciation to Prof. Hiroaki Ohno, for his kind guidance, constructive discussions, and constant encouragement throughout this study.

The author would like to express his sincere and heartfelt gratitude to Dr. Shinya Oishi and Dr. Shinsuke Inuki (Kyoto University) for their valuable discussions and positive encouragement.

The author would also like to express his sincere appreciation to Prof. Yuji Mori and Dr. Takeo Sakai (Faculty of Pharmacy, Meijo University) for their useful advice and encouragement during this study.

The author is profoundly grateful to Prof. Masanobu Uchiyama, Dr. Tatsuo Saito, Dr. Chao Wang (The University of Tokyo), Prof. Masahiro Oka, Dr. Yoichi Miyamoto, Ms. Mayumi Otani (National Institutes of Biomedical Innovation, Health and Nutrition) for their professional guidance as well as technical supports.

The author would like thank Prof. Jieping Zhu (Institute of Chemical Sciences and Engineering, Ecole Polytechnique Fédérale deLausanne, EPFL, Switzerland) for providing him with an opportunity to study total synthesis of alkaloid at EPFL, Switzerland.

The supports and advices from Prof. Yoshiji Takemoto (Graduate School of Pharmaceutical Sciences, Kyoto University) and Prof. Kiyosei Takasu (Graduate School of Pharmaceutical Sciences, Kyoto University) are greatly appreciated.

The author also wishes to express his gratitude to all the colleagues in the Department of Bioorganic Medicinal Chemistry/Department of Chemogenomics (Graduate School of Pharmaceutical Sciences, Kyoto University) for their valuable comments and for their assistance and cooperation in various experiments. The author is especially grateful to Ms. Yuka Matsuda, Mr. Yuiki Kawada and Mr. Hiroshi Kumagai for the assistance with his experiments.

The author would like to thank the Japan Society for the Promotion of Science (JSPS) and Nagai Memorial Research Scholarship from the Pharmaceutical Society of Japan for financial support and all the staffs at the Elemental Analysis Center, Kyoto University.

Contents

Chapter 1
Introduction

Abstract This thesis focuses on the efficient syntheses of indole alkaloids based on gold-catalyzed cascade cyclizations. Total syntheses of these natural products have been achieved using two strategies for the construction of pyrrolocarbazole cores via gold-catalyzed reaction of conjugated diyne or ynamide. This chapter provides an overview of biosynthesis and previous total synthesis of dictyodendrins and vindorosine, then introduced the gold-catalyzed reaction of alkynes.

1.1 Pyrrolocarbazoles

Nitrogen-containing heterocyclic scaffolds are found in many biologically active compounds, including natural and synthetic products. Carbazoles are well known for their pharmacological activities, including anticancer, antioxidant, anti-inflammatory, antibacterial, and antitumor activities. Pyrrolocarbazoles, which are tetracyclic compounds composed of a carbazole fused with a pyrrole ring, are of great interest to organic and medicinal chemists because these structures exist in a variety of biologically active compounds. Pyrrolocarbazoles are categorized by the position of the carbazole–pyrrole ring junction, as well as the relative orientation of the pyrrole nitrogen atom with respect to the carbazole moiety (Fig. 1.1) [1, 2]. Pyrrolocarbazoles with further ring fusion(s), such as indolocarbazoles, are important structural motifs.

Many natural products having a pyrrolocarbazole scaffold possess important biological activity (Fig. 1.2). For example, staurosporine (**1**), an indolo[2,3-*a*]carbazole isolated from a bacterium, is a nanomolar inhibitor of several protein kinases [3, 4]. Vindorosine (**2**) [5, 6], vinblastine (**3**), and vincristine (**4**), which contain a pyrrolo[2,3-*d*]carbazole scaffold, are widely used as anticancer drugs [7–9]. The pyrrolo[2,3-*c*]carbazoles, dictyodendrins A (**5**) and B (**6**), have been reported to exhibit several biological activities, including telomerase inhibition (vide infra). Thus, pyrrolocarbazoles are a pharmacologically important class of natural products.

© The Editor(s) (if applicable) and The Author(s), under exclusive license to Springer   1
Nature Singapore Pte Ltd. 2020
J. Matsuoka, *Total Synthesis of Indole Alkaloids*, Springer Theses,
https://doi.org/10.1007/978-981-15-8652-1_1

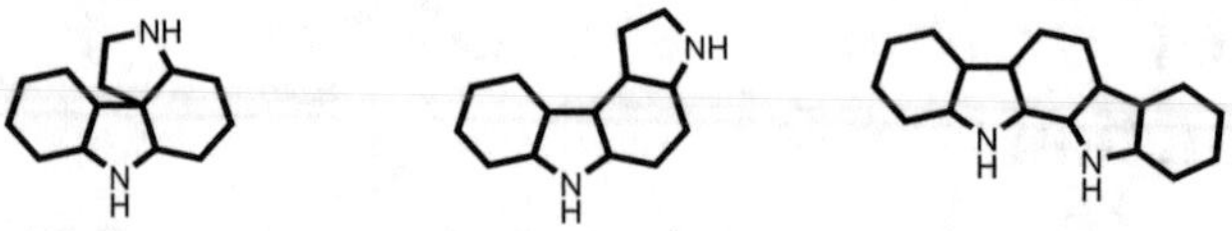

pyrrolo[2,3-*d*]carbazole   pyrrolo[2,3-*c*]carbazole   indolo[2,3-*a*]carbazole

Fig. 1.1 Pyrrolocarbazole scaffolds

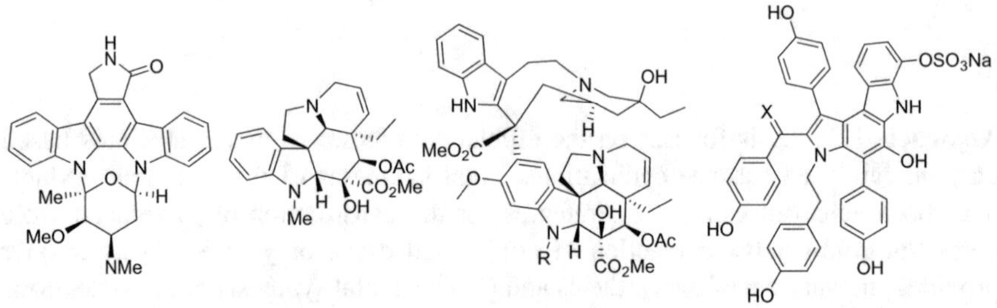

staurosporine (**1**)   vindorosine (**2**)   vinblastine (**3**): R = CH₃   dictyodendrin A (**5**):
                                              vincristine (**4**): R = CHO   X = (CO₂Me, H)
                                                                        dictyodendrin B (**6**):
                                                                      X = O

Fig. 1.2 Pharmacologically important natural products containing a pyrrolocarbazole scaffold

1.1.1 Dictyodendrins

Dictyodendrins A–E (**5–9**, Figs. 1.2 and 1.3) were initially isolated by Fuse-tani and co-workers from the Japanese marine sponge *Dictyodendrilla verongi-formis* in 2003 [10]. Dictyodendrins F–I (**10–13**) were isolated by Capon and co-workers in 2012 from the southern Australian marine sponge *Ianthella* sp. [11]. The key structural characteristic of the dictyodendrins is a pyrrolo[2,3-*c*]carbazole

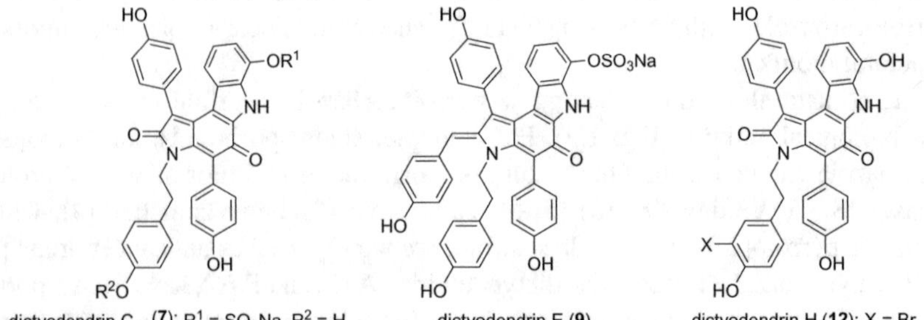

dictyodendrin C (**7**): R¹ = SO₃Na, R² = H   dictyodendrin E (**9**)   dictyodendrin H (**12**): X = Br
dictyodendrin D (**8**): R¹ = SO₃Na, R² = SO₃Na   dictyodendrin I (**13**): X = I
dictyodendrin F (**10**): R¹ = H, R² = H
dictyodendrin G (**11**): R¹ = Me, R² = H

Fig. 1.3 Structures of the dictyodendrins

core bearing oxygenated functional groups. Dictyodendrins show inhibitory activity toward telomerase and β-site amyloid-cleaving enzyme 1 (BACE1).

The biosynthesis of dictyodendrins was suggested by Ready in 2017 (Scheme 1.1) [12]. Ready proposed that dictyodendrins are biosynthesized from tryptophan (**14**) and tyrosine (**15**). The oxidative coupling and the subsequent Paal–Knorr-type condensation of the resulting diketone **16** with a second molecule of tyrosine (**15**) produce fully substituted pyrrole **17**. An oxidative decarboxylation of pyrrole **17** then gives maleimide **18**, which can be transformed to carboxylic acid **19** via an oxidative aldol-type condensation. Subsequently, decarboxylation and cyclization afford pyrrole[2,3-*c*]carbazole **20**. A series of dictyodendrins can then be produced via further transformations of pyrrolocarbazole **20**.

The reported total syntheses of dictyodendrins can be categorized into two main synthetic strategies [12]: (1) benzene ring formation from substituted indoles (Fürstner [13–15], Ready [16], Ishibashi [17, 18], and Yamaguchi/Itami/Davies [19]) and (2) pyrrole ring formation from substituted indoles (Tokuyama [20, 21], Jia, and Gaunt [24]) as shown in Fig. 1.4.

(1) Benzene ring formation from substituted indoles

The Fürstner group reported the first total synthesis of dictyodendrins in 2005–2006 (Scheme 1.2) [13–15]. This group employed 6π-electron cyclization of compound **23**, which was derived from chalcone **21** readily prepared via TiCl3-mediated reductive cyclization of compound **22**, for formation of the pyrrolocarbazole core in compound **24**. Introduction of acyl and sulfate groups at the C2 and

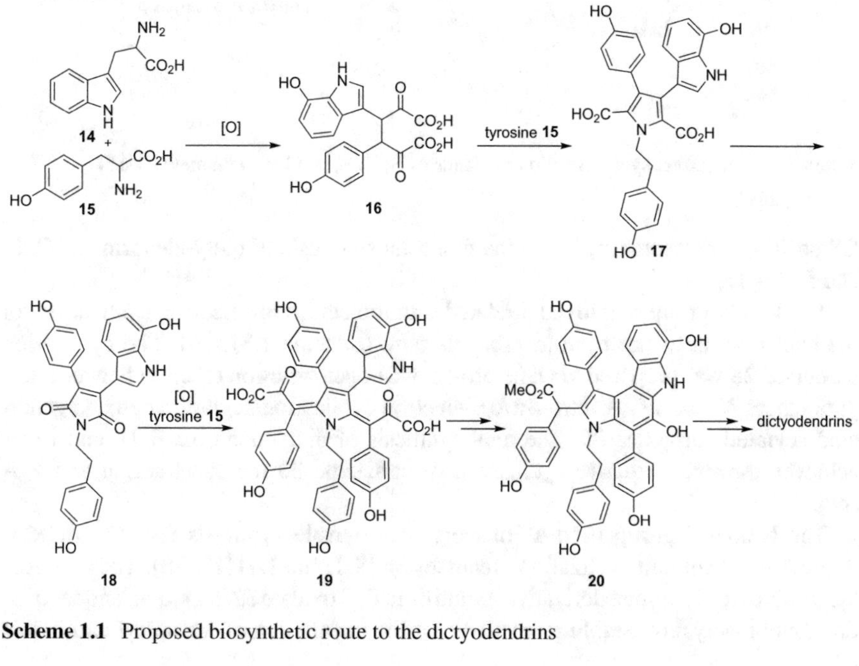

Scheme 1.1 Proposed biosynthetic route to the dictyodendrins

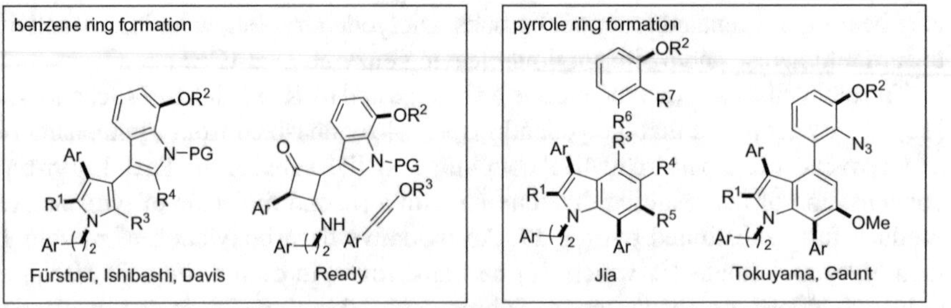

Fig. 1.4 Two main synthetic strategies for the dictyodendrin scaffold

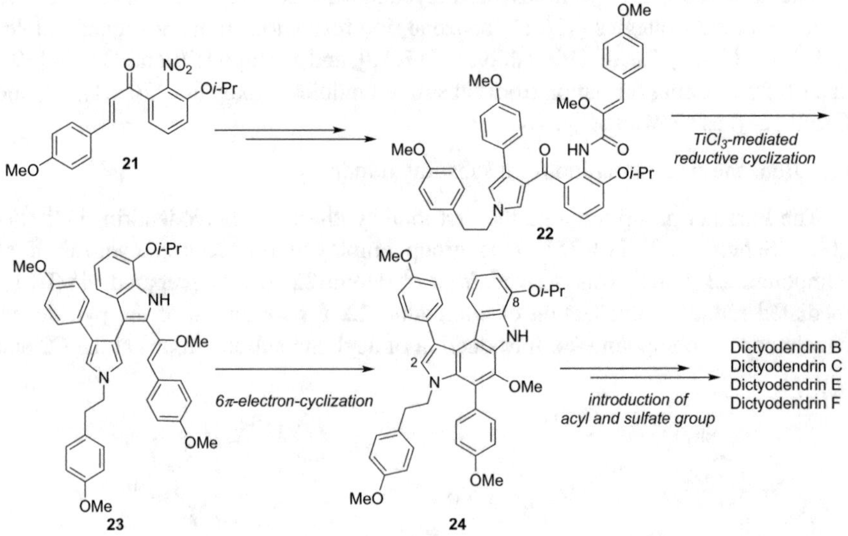

Scheme 1.2 First total synthesis of dictyodendrins B, C, E, and F by Fürstner [13–15]

C8 positions, respectively, led to the first total synthesis of dictyodendrins B, C, E, and F [13–15].

The Ready group employed a related intramolecular 6π-electron cyclization for the construction of the pyrrolocarbazole core (Scheme 1.3) [16]. The cyclization precursor **28** was prepared via a hetero-[2 + 2]-cycloaddition reaction between aryl ynol ethers **25** and **26**. A retro-4π/6π-electron cyclization/acylation cascade generated acylated carbazole **30**. The total synthesis of dictyodendrins F, H, and I was achieved through oxidative cyclization of carbazole **30** for construction of the A ring.

The Ishibashi group used a Hinsberg-type pyrrole synthesis for the synthesis of pyrrole-substituted cyclization precursor **36** (Scheme 1.4) [17, 18]. Thus, pyrrole synthesis using tyramine derivative **33** and dimethyl oxalate **32**, triflation, and sequential Suzuki–Miyaura coupling reactions for the installation of the anisyl and indole

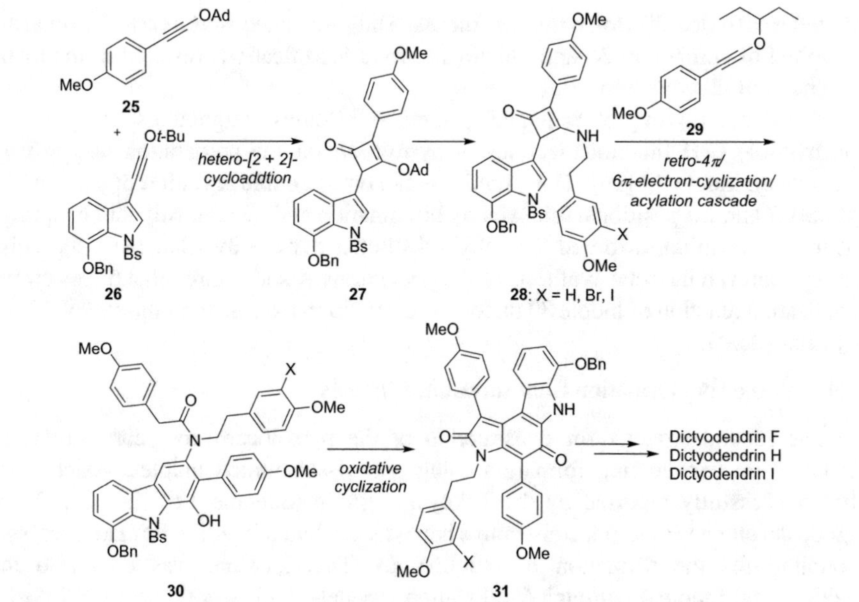

Scheme 1.3 First total synthesis of dictyodendrins H and I by Ready [16]

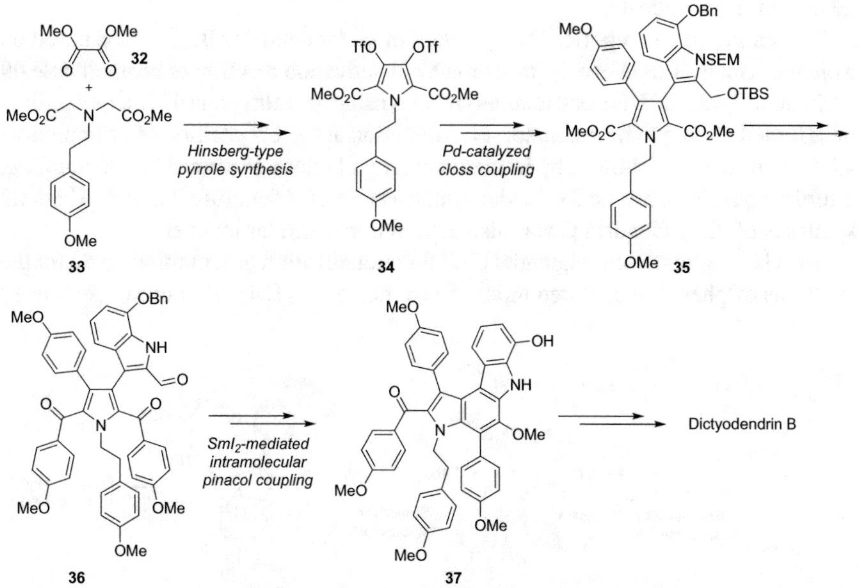

Scheme 1.4 Formal total synthesis of dictyodendrin B by Ishibashi [17, 18]

moieties afforded diketo-pyrrolylindole **36**. SmI$_2$-mediated intramolecular pinacol coupling of compound **36** and functional group modifications resulted in the total synthesis of dictyodendrin B.

The research group of Yamaguchi, Itami, and Davies designed a strategy based on multiple C-H functionalizations of pyrrole **38** for the preparation of pyrrole-substituted indole **39** [19]. Thus, rhodium-catalyzed double arylation of pyrrole **38** at the C2 and C5 positions, followed by bromination and Suzuki–Miyaura coupling at the C4 position, afforded the fully substituted pyrrole **39** (Scheme 1.5). This group achieved the total synthesis of dictyodendrins A and F through a 6π-electron cyclization reaction of indole **39** under basic conditions to construct the pyrrolo[2,3-c]carbazole.

(2) Pyrrole ring formation from substituted indoles

The second strategy for construction of the pyrrolocarbazole core structure is based on pyrrole ring formation using phenyl-substituted indoles, which was first successfully reported by the Tokuyama group (Scheme 1.6) [20, 21]. This group developed a one-pot consecutive benzyne-mediated indoline formation/cross-coupling for the formation of indoline **43**. This indoline was converted to azidophenyl-indole **45** through N-alkylation, Friedel–Crafts reaction at the C2 position, and Suzuki–Miyaura coupling at the C4 position. The total synthesis of dictyo-dendrins A–E was accomplished by pyrrole ring formation through a nitrene species generated from azide **45**.

The Jia group has reported the synthesis of dictyodendrins B, C, and E based on a one-pot Buchwald–Hartwig amination/C–H activation reaction of bromoindole **49** (Scheme 1.7) [22, 23]. Larock indole synthesis using aryl alkynone **47** and iodoaniline **48** afforded the requisite precursor **49**. Amination at the C5 position of bromoindole **49** with aniline **50**, followed by intramolecular palladium-catalyzed biaryl coupling, afforded pyrrolocarbazole **51** for the synthesis of dictyodendrins B and E. The total synthesis of dictyodendrin C was also achieved in a similar manner.

The Gaunt group used sequential C–H functionalization of bromoindole **52** for the synthesis of phenyl-substituted indole **53** (Scheme 1.8) [24]. This group completed

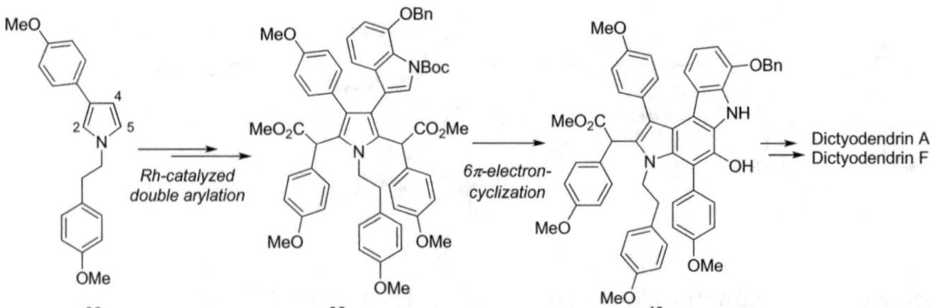

Scheme 1.5 Total synthesis of dictyodendrins A and F by Yamaguchi, Itami, and Davis [19]

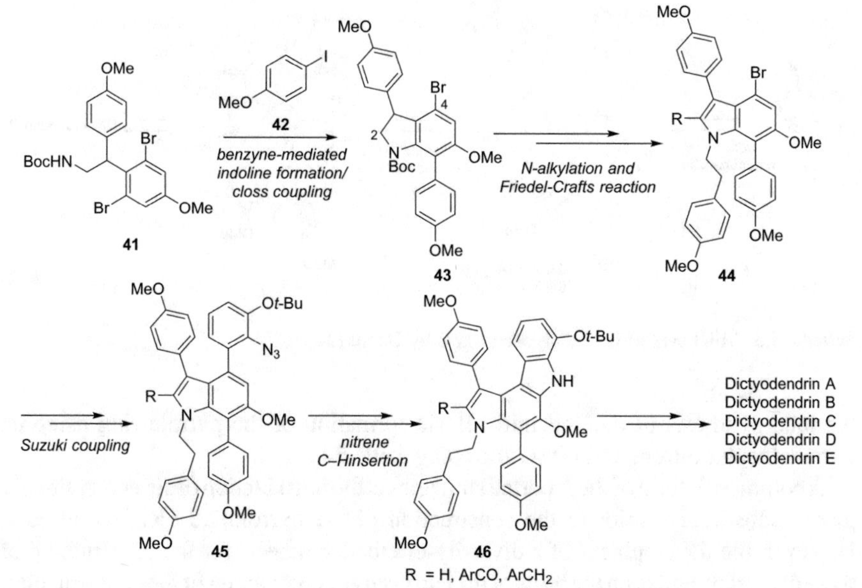

Scheme 1.6 Total synthesis of dictyodendrins A–E by Tokuyama [20, 21]

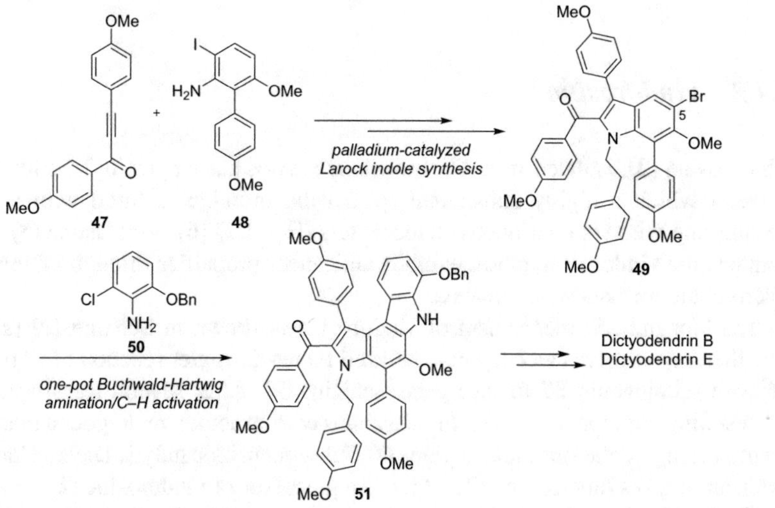

Scheme 1.7 Total synthesis of dictyodendrins B and E by Jia [22, 23]

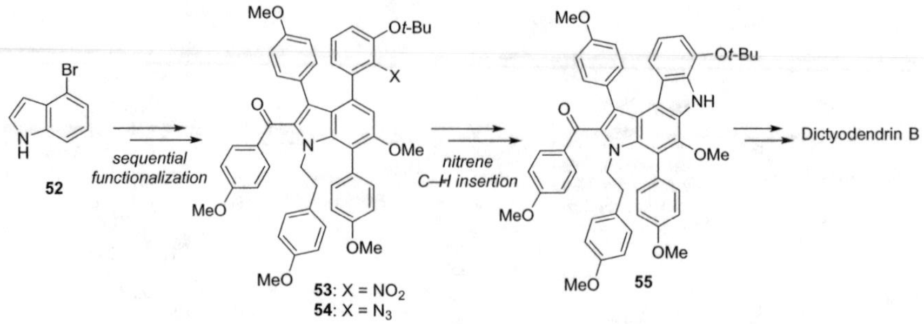

Scheme 1.8 Total synthesis of dictyodendrin B by Gaunt [24]

the total synthesis of dictyodendrin B via formation of the pyrrole ring using an intramolecular nitrene C–H insertion using azide **54**.

A common feature of the reported strategies is the introduction of several optimally placed substituents prior to the construction of the pyrrolo[2,3-*c*]carbazole core. However, the development of a diversity-oriented synthesis for the construction of these natural products on the basis of the early-stage construction of the core structure, followed by the introduction of the different substituents, would be more amenable to medicinal applications.

1.1.2 Vindorosine

Vindorosine (**2**), isolated from *Catharanthus roseus*, has a pyrrolo[2,3-*d*]carbazole core, in which a highly substituted spirocyclic indoline is fused with a cyclo-hexane ring with six continuous stereocenters (Fig. 1.2) [6]. Vinblastine (**3**), which contains the vindorosine motif, exhibits anticancer properties through inhibition of microtubule formation and mitosis.

The biosynthetic mechanism of vindorosine is shown in Scheme 1.9 [25–27]. The first step involves an enzyme-catalyzed Pictet–Spengler reaction of tryptamine **56** with secologanin **57** to give geissoschizine **58**. After several transformations, the resulting compound **59**, with the strychnos core structure, undergoes a rearrange-ment and fragmentation to afford triene **60**. Subsequent biocatalytic Diels–Alder-type cyclization gives tabersonine **61**, which is a precursor of vindorosine (**2**).

The first total synthesis of vindorosine was reported by the Büchi group in 1971 (Scheme 1.10) [28]. The key step involves an intramolecular Robinson-type annu-lation of enamido ketone **64** to form tetracyclic indoline **65**, which is called the Büchi ketone [29, 30]. The pentacyclic compound **67** was obtained by Michael addi-tion of compound **65** to acrolein **66** and a subsequent intramolecular aldol reaction. Vindorosine (**2**) was then obtained by stereoselective introduction of several func-tional groups to compound **67**. This total synthesis, although racemic, provides one of the simplest, shortest, and most efficient routes to vindorosine to date.

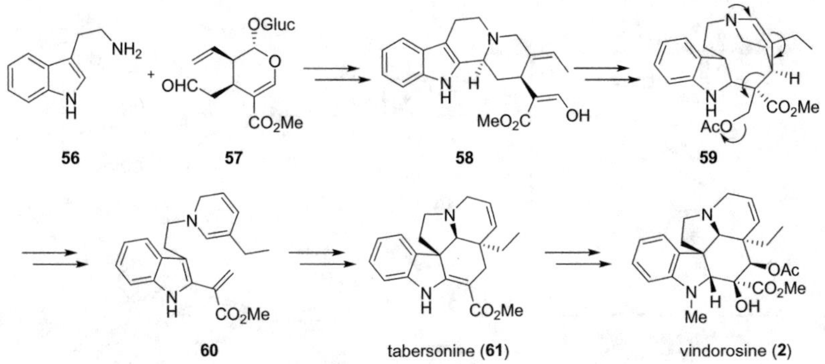

Scheme 1.9 Proposed biosynthesis of vindorosine

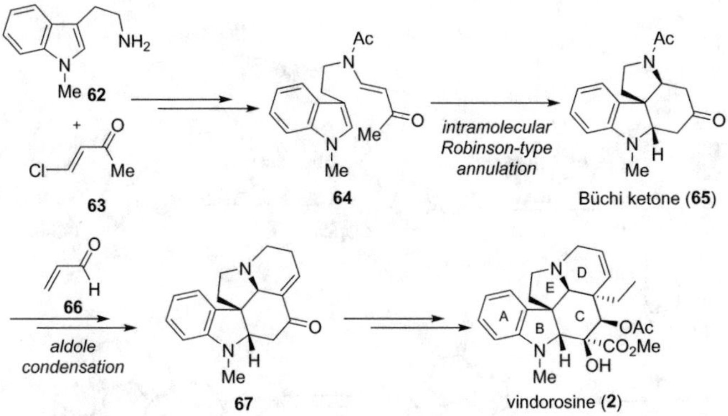

Scheme 1.10 First total synthesis of vindorosine by Büchi in 1971 [28]

The Kuehne group reported the first asymmetric total synthesis of vindorosine in 1986 (Scheme 1.11) [31]. Following the biosynthetic hypothesis, pentacyclic compound **72** was obtained through a one-pot conversion of fused-indole derivative **68** via condensation–sigmatropic rearrangement using hemiacetal **69** as the coupling partner.

The Boger group reported the asymmetric total synthesis of vindorosine (**2**) via an intramolecular [4 + 2] and [3 + 2] cycloaddition cascade of oxadiazole **78** (Scheme 1.12) [32, 33]. This cascade reaction formed three rings and four C–C bonds in a single step, leading to formation of the pentacyclic compound **79** with all six stereocenters. This group also completed the total synthesis of vinblastine (**3**) and vincristine (**4**) in a similar manner. Structure–activity relationship studies of vindorosine derivatives were also conducted.

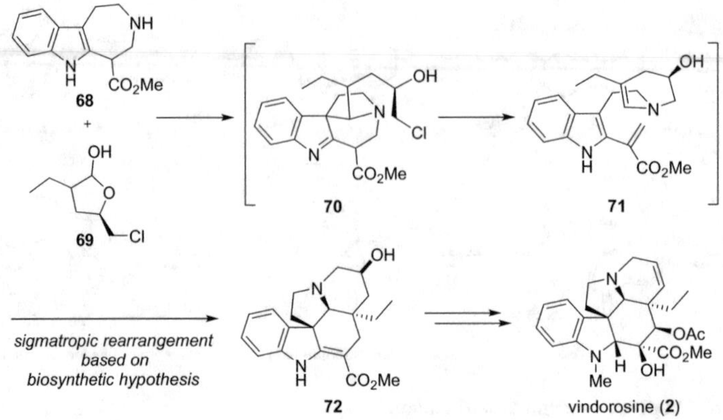

Scheme 1.11 First asymmetric total synthesis of vindorosine by Kuehne in 1986 [31]

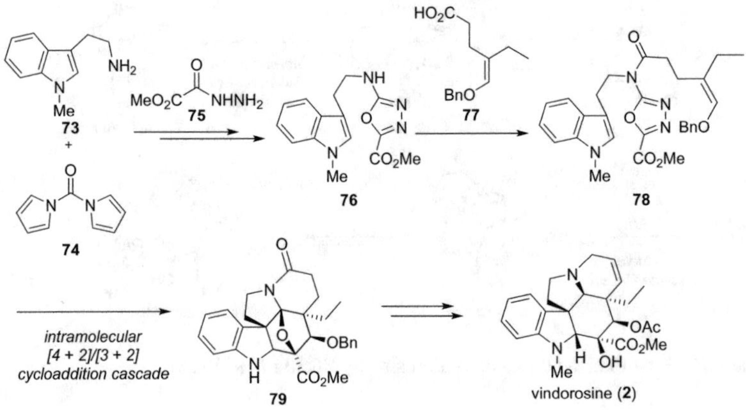

Scheme 1.12 Asymmetric total synthesis of vindorosine by Boger in 2006 [32, 33]

1.2 Gold-Catalyzed Reactions of Alkynes

Over the last decade, the use of gold complexes for the electrophilic activation of
alkynes has become a powerful tool for increasing molecular complexity in an atom-
economical fashion [34–36]. In these gold-catalyzed reactions, vinyl–gold intermediates are formed by nucleophilic attack on activated alkynes (Scheme 1.13). Subsequent electrophilic attack of the vinyl–gold intermediates can be classified into two
types: deauration by the electrophile leading to formation of alkenes bearing two
newly introduced substituents (Eq. 1.1) and formation of gold carbenoids, which
can undergo further transformations, including C–H insertion and cyclopropanation (Eq. 1.2). Various types of alkynes can be employed for these transformations,
including ynamides, enynes, and diynes.

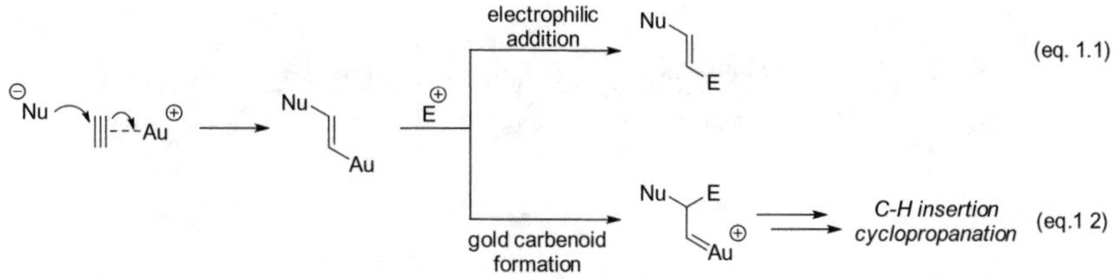

(eq. 1.1)

(eq. 1.2)

Scheme 1.13 Gold-catalyzed reactions of alkynes

1.2.1 Gold-Catalyzed Reactions of Ynamides

Ynamides are alkynes that have the alkyne functionality directly attached to a nitrogen atom bearing an electron-withdrawing amido group (Scheme 1.14) [37–41]. The reactions of ynamides provide access to nitrogen-containing compounds, as well as to important structural motifs existing in natural and medicinal products. In terms of the reactivity of ynamides, the electron-donating nitrogen atom strongly polarizes the C–C triple bond, leading to high chemo- and regioselectivity in the transformations of the ynamides.

Gold-catalyzed cascade reactions of ynamides provide straightforward access to polycyclic compounds containing a nitrogen atom. For example, Cossy has reported a gold(I)-catalyzed cyclization of ene–ynamide **80** to form aza-bicycle **82** (Scheme 1.15) [42]. This cycloisomerization is believed to proceed through an oxygen-assisted hydride shift from gold carbenoid species **81**.

During the course of the author's study, Yang and Gong reported an efficient synthesis of spirocyclic indoline **85** through a gold(I)-catalyzed intramolecular cascade cyclization (Scheme 1.16) [43]. Activation of indolyl-ynamide **83** by coordination of a gold catalyst promoted the formation of the iminium intermediate **84**,

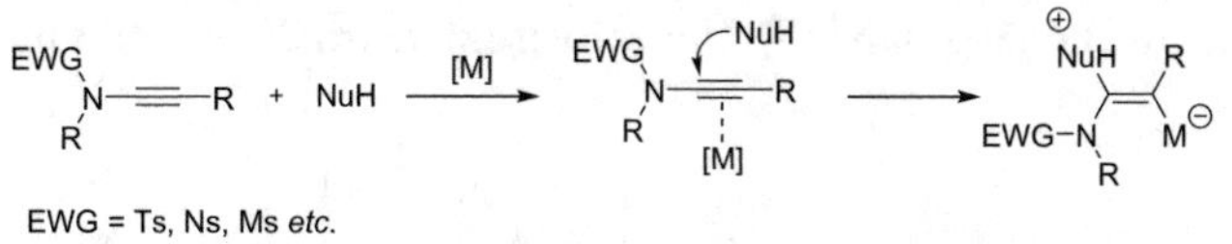

EWG = Ts, Ns, Ms *etc.*

Scheme 1.14 Transition metal-catalyzed reaction of ynamides

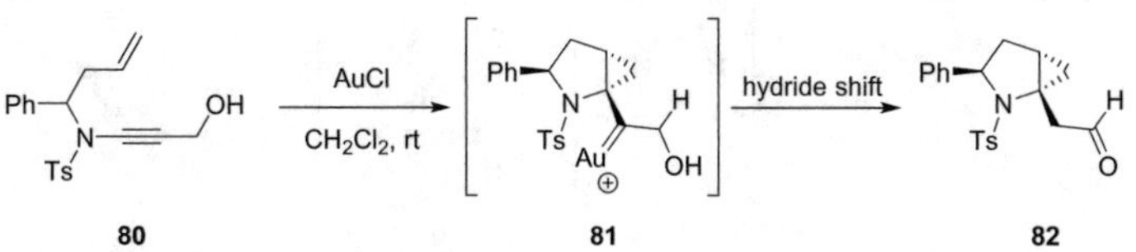

Scheme 1.15 Gold-catalyzed construction of nitrogen-containing bicyclic compounds

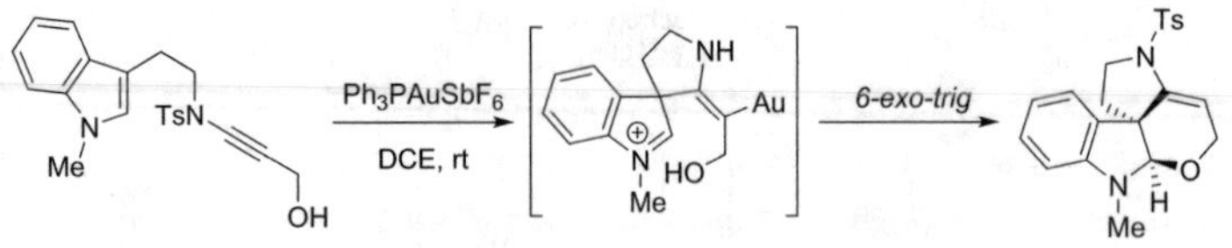

Scheme 1.16 Gold(I)-catalyzed intramolecular cascade cyclization of an indolyl-ynamide

which was converted to the tetracyclic indoline **85** by intramolecular aminal formation. However, the gold-catalyzed cascade cyclization of ynamides via contiguous formation of a C–C bond was still unknown.

1.2.2   Gold-Catalyzed Cyclization of Diynes

Conjugated diynes (1,3-diynes), which possess two C–C triple bonds, are useful building blocks for the synthesis of cyclic compounds and are also found in natural products and bioactive compounds [44, 45]. The gold-catalyzed reactions of conjugated alkynes are relatively undeveloped compared with those of isolated alkynes. The general reaction modes for the intermolecular nucleophilic reactions of conjugated diynes are shown in Scheme 1.17. Addition of nucleophilic reagent **87** or alkene **88** to activated alkyne **86** leads to formation of enynes **89** or **90**, which deliver cyclization products **91** or **92**, respectively, by intramolecular nucleophilic attack.

Skrydstrup's group has reported an innovative synthesis of substituted pyrroles or furans by gold(I)-catalyzed annulation of diyne **93** with aniline **94** (Scheme 1.18, Eq. 1.3) [46]. This reaction proceeds through a double hydroamination cascade of two alkynes to form disubstituted pyrrole **95**. Banwell's group has demonstrated a gold(I)-catalyzed cascade cyclization of conjugated diyne **96** bearing a urea group as a dual nucleophilic functional group [47]. Under microwave irradiation, this reaction gives

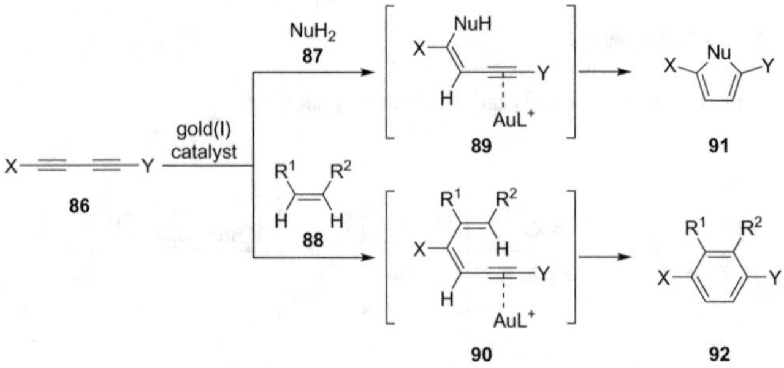

Scheme 1.17 Gold(I)-catalyzed tandem cyclization of conjugated diynes

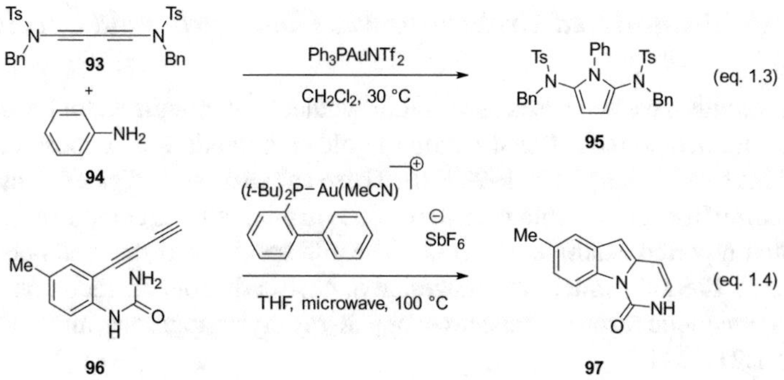

Scheme 1.18 Gold(I)-catalyzed construction of heterocycles via conjugated diynes

tricyclic product **97** through gold(I)-catalyzed 5-*endo-dig* and 6-*endo-dig* cyclization steps (Eq. 1.4).

Recently, the author's group has reported the intermolecular reaction of conjugated diynes **98** and pyrrole **99** for the synthesis of 4,7-disubstituted indoles **101** (Scheme 1.19) [48]. This reaction proceeds through a double hydroarylation cascade involving the two alkyne groups to form a disubstituted benzene ring. Thus, conjugated alkynes can be considered a promising substrate for gold-catalyzed cascade cyclization.

Unconjugated diynes are also useful substrates in gold-catalyzed reactions. The author's group has previously developed a gold(I)-catalyzed cyclization of diyne **102** containing an aniline moiety (Scheme 1.20) [49]. This reaction directly produced aryl-annulated[*a*]carbazoles **104** by intramolecular cascade hydroamination/cycloisomerization.

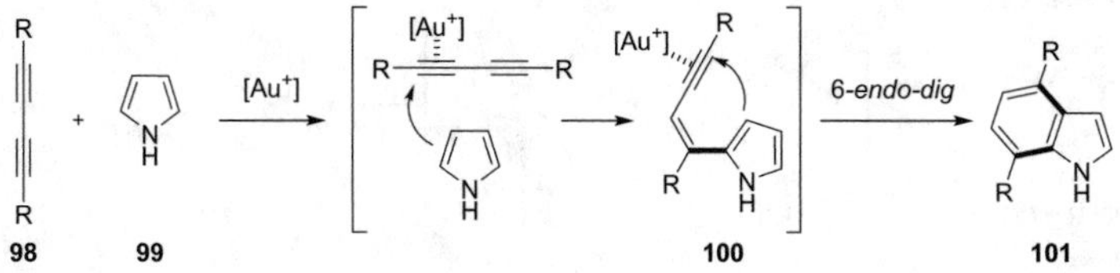

Scheme 1.19 Gold(I)-catalyzed cascade annulation reported by the author's group

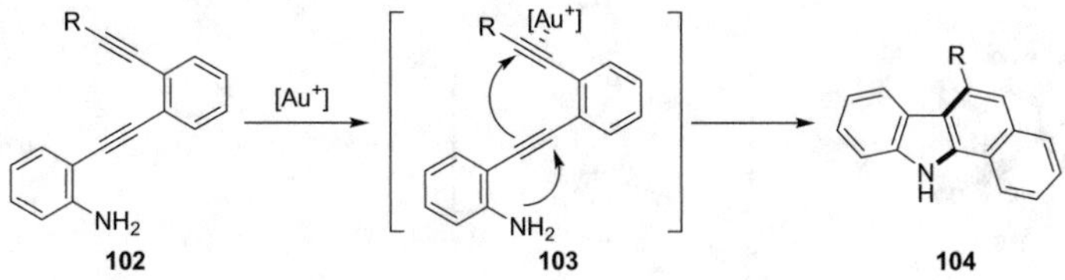

Scheme 1.20 Gold(I)-catalyzed cascade annulation reported by the author's group

1.2.3   Gold-Catalyzed Cyclization via Gold Carbenoid Species

Gold carbenoids have been used as versatile synthetic intermediates for the construction of cyclic compounds. The structure of gold carbenoids was proposed based on
calculations by Goddard III in 1994 [50]. These calculations indicated considerable
back-bonding from gold(I) into vacant p orbitals in gold–alkyne complexes (Fig. 1.5).

The first reported example of a gold carbenoid species was the metal-complexed
gold species **106/107**, which was suggested to have carbenoid character on the basis
of nuclear magnetic resonance spectroscopy, X-ray crystallography, and calculations
(Scheme 1.21) [51].

Reactions using gold carbenoid species were reported nearly simultaneously by
the groups of Echavarren [52], Fürstner [53], and Toste [54] in their studies on gold(I)-
catalyzed cycloisomerizations of enynes (Scheme 1.22, Eqs. 1.5–1.7). Echavarren
demonstrated that the gold(I)-catalyzed reaction of 1,6-enyne **108** generated gold
carbenoid intermediate **109** by nucleophilic attack of the alkene on the activated

Fig. 1.5 Proposed gold
carbenoid structure

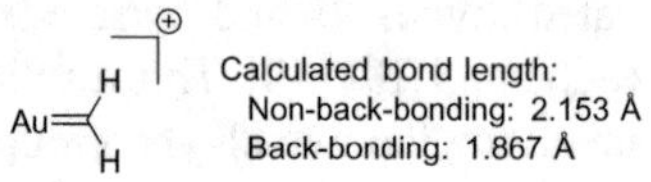

Calculated bond length:
Non-back-bonding: 2.153 Å
Back-bonding: 1.867 Å

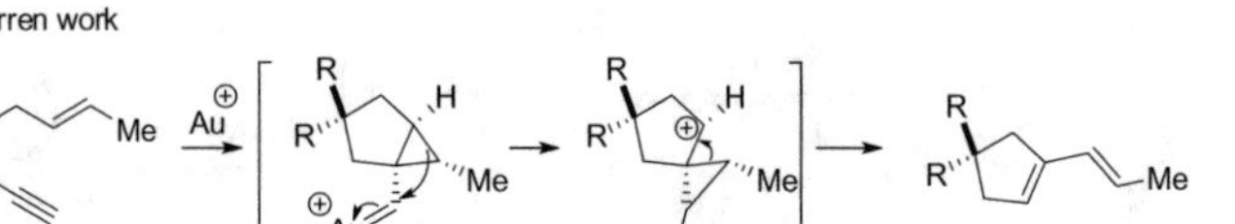

Scheme 1.21 First example of a gold carbenoid complex

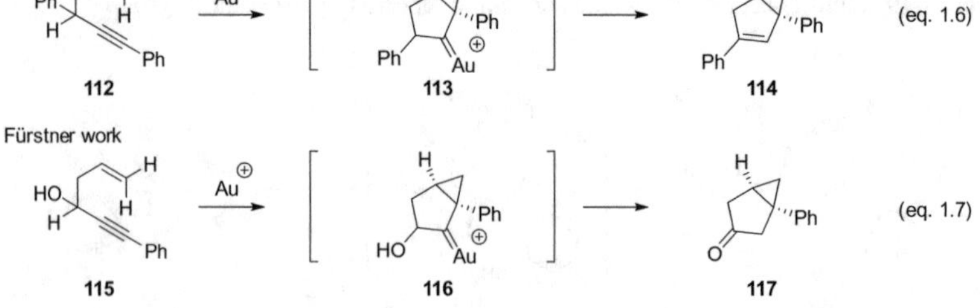

Scheme 1.22 Gold(I)-catalyzed cycloisomerization-type reaction of enynes

alkyne, and cyclopropanation. Cleavage of the C–C bond with skeletal rearrangement of compound **110** accompanying deauration produced alkene-substituted cyclopentene **111** (Eq. 1.5). Fürstner and Toste reported that the reaction of 1,5-enynes **112** and **115** generated cyclic gold carbenoids **113** and **116**, which underwent 1,2-hydride shift onto the gold carbenoid to give bicyclic compounds **114** and **117** (Eqs. 1.6 and 1.7).

1.2.4   Gold-Catalyzed Acetylenic Schmidt Reaction via Gold Carbenoid Species

Gold carbenoids can be generated by an acetylenic Schmidt reaction as shown in Scheme 1.23 [36]. Gold-catalyzed nucleophilic addition to the pendant alkyne from an ylide-type nucleophilic functional group, followed by elimination of a leaving group, gives carbenoid species **120** (Scheme 1.23, Eq. 1.8). In contrast, the reaction of compound **121** bearing an ylide moiety with a nucleophilic group at the terminus affords acyclic carbenoid species **123** through a ring-opening step (Eq. 1.9).

Toste and co-workers have reported pioneering work on gold(I)-catalyzed intramolecular acetylenic Schmidt reactions (Scheme 1.24) [55]. The gold-catalyzed

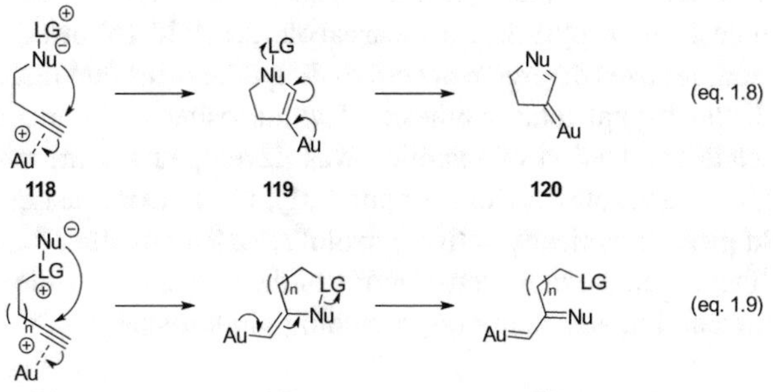

Scheme 1.23   Generation of gold carbenoids from ylide-tethered alkynes

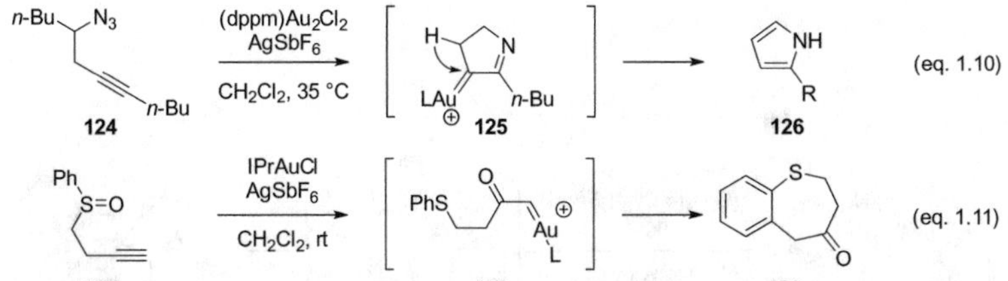

Scheme 1.24   Gold(I)-catalyzed intramolecular Schmidt reaction and sulfoxide rearrangement

reaction of azido-alkyne **124** led to the formation of cyclic gold carbenoid intermediate **125**, which was trapped by 1,2-hydride shift and subsequent tautomerization to
form pyrrole **126** (Eq. 1.10). This group also disclosed that the gold(I)-catalyzed reaction of sulfoxide **127** provided acyclic carbenoid intermediate **128**, which underwent
an intramolecular C–H insertion of a phenyl group to afford benzothiepine derivative
129 (Eq. 1.11) [56]. Following these pioneering works, the author's group disclosed
the development of a method for the construction of indoloquinoline compounds
by the gold-catalyzed cascade cyclization of (azido)ynamides (Scheme 1.25) [57].
The reaction of (azido)ynamide **130** with a gold catalyst led to the formation of α-
amidino gold carbenoid **131**, which was transformed to indoloquinoline **132** through
an intramolecular trapping reaction with an alkene. These reactions clearly show
the high potential of this strategy for construction of complex heterocycle scaffolds. However, the use of diynes in acetylenic Schmidt reactions, as well as in
intermolecular trapping with pyrrole-type nucleophiles, is unknown.

In this thesis, the total synthesis of dictyodendrins and vindorosine based on
gold-catalyzed cascade cyclizations is described. An efficient total synthesis of
these natural products was developed using two strategies for the construction of
the pyrrolocarbazole core structures on the basis of the gold-catalyzed reactions of
diynes or ynamides.

In Chap. 2, the total synthesis of dictyodendrins is described. An acetylenic
Schmidt reaction of azido-diynes with a pyrrole derivative was developed for
the direct construction of the pyrrolocarbazole core. This strategy, based on
the facile synthesis of a pyrrole[2,3-*c*]carbazole scaffold followed by late-stage
functionalization, realizes divergent access to dictyodendrins and their derivatives.

In Chap. 3, the formal total synthesis of vindorosine is described. A gold(I)-
catalyzed cascade cyclization of ynamide was developed for the construction of
the pyrrolo[2,3-*d*]carbazole scaffold. Importantly, the reaction using a chiral gold
complex could provide optically active pyrrolo[2,3-*d*]carbazole. This newly developed strategy facilitated the rapid construction of the pyrrolocarbazole core structure
of Aspidosperma and related alkaloids, including vindorosine.

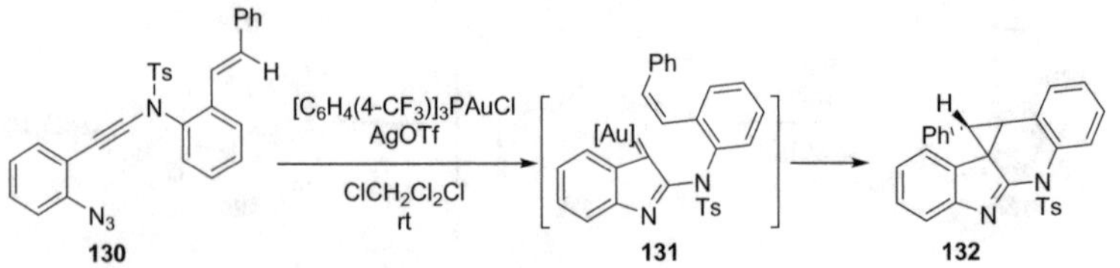

Scheme 1.25 Gold(I)-catalyzed intramolecular Schmidt reaction of (azido)ynamides

References

1. Tahlan S, Kumar S, Narasimhan B (2019) BMC Chem 13:1–21
2. Issa S, Prandina A, Bedel N, Rongved P, Yous S, Le Borgne M, Bouaziz Z (2019) J Enzyme Inhib Med Chem 34:1321–1346
3. Ōmura S, Iwai Y, Hirano A, Nakagawa A, Awaya J, Tsuchiya H, Takahashi Y, Masuma R (1977) J Antibiot 30:275–282
4. Ōmura S, Sasaki Y, Iwai Y, Takeshima H (1995) J Antibiot 48:535–548
5. Gorman M, Neuss N, Biemann K (1962) J Am Chem Soc 84:1058–1059
6. William JM, Lipscomb WN (1965) J Am Chem Soc 87:4963–4964
7. Ishikawa H, Colby DA, Seto S, Va P, Tam A, Kakei H, Rayl TJ, Hwang I, Boger DL (2009) J Am Chem Soc 131:4904–4916
8. Schleicher KD, Sasaki Y, Tam A, Kato D, Duncan KK, Boger DL (2013) J Med Chem 56:483–495
9. Sears JE, Boger DL (2015) Acc Chem Res 48:653–662
10. Warabi K, Matsunaga S, van Soest RWM, Fusetani N (2003) J Org Chem 68:2765–2770
11. Zhang H, Conte MM, Khalil Z, Huang X-C, Capon RJ (2012) RSC Adv 2:4209–4214
12. Zhang W, Ready JM (2017) Nat Prod Rep 34:1010–1034
13. Fürstner A, Domostoj MM, Scheiper B (2005) J Am Chem Soc 127:11620–11621
14. Fürstner A, Domostoj MM, Scheiper B (2006) J Am Chem Soc 128:8087–8094
15. Buchgraber P, Domostoj MM, Scheiper B, Wirtz C, Mynott R, Rust J, Fürstner A (2009) Tetrahedron 65:6519–6534
16. Zhang W, Ready JM (2016) J Am Chem Soc 138:10684–10692
17. Hirao S, Sugiyama Y, Iwao M, Ishibashi F (2009) Biosci Biotechnol Biochem 73:1764–1772
18. Hirao S, Yoshinaga Y, Iwao M, Ishibashi F (2010) Tetrahedron Lett 51:533–536
19. Yamaguchi AD, Chepiga KM, Yamaguchi J, Itami K, Davies HML (2015) J Am Chem Soc 137:644–647
20. Okano K, Fujiwara H, Noji T, Fukuyama T, Tokuyama H (2010) Angew Chem Int Ed 49:5925–5929
21. Tokuyama H, Okano K, Fujiwara H, Noji T, Fukuyama T (2011) Chem Asian J 6:560–572
22. Liang J, Hu W, Tao P, Jia Y (2013) J Org Chem 78:5810–5815
23. Tao P, Liang J, Jia Y (2014) Eur J Org Chem 2014:5735–5748
24. Pitts AK, O'Hara F, Snell RH, Gaunt MJ (2015) Angew Chem Int Ed 54:5451–5455
25. O'Connor SE, Maresh JJ (2006) Nat Prod Rep 23:532–547
26. Tatsis EC, Carqueijeiro I (2017) Dugé De Bernonville T, Franke J, Dang TTT, Oudin A, Lanoue A, Lafontaine F, Stavrinides AK, Clastre M, Courdavailt V, O'Connor SE. Nat Commun 8:1–9
27. Saya JM, Ruijter E, Orru RVA (2019) Chem Eur J 25:8916–8935
28. Matsumoto KE, Büchi G, Nishimura H (1971) J Am Chem Soc 93:3299–3301
29. Heureux N, Wouters J, Markó IE (2005) Org Lett 7:5245–5248
30. Wang Y, Lin J, Wang X, Wang G, Zhang X, Yao B, Zhao Y, Yu P, Lin B, Liu Y, Cheng M (2018) Chem Eur J 24:4026–4032
31. Kuehne ME, Bornmann WG, Earley WG, Istvan Marko I (1986) M. J Org Chem 51:2913–2927
32. Elliott GI, Velcicky J, Ishikawa H, Li YK, Boger DL (2006) Angew Chem Int Ed 45:620–622
33. Sasaki Y, Kato D, Boger DL (2010) J Am Chem Soc 132:13533–13544
34. Gorin DJ, Toste FD (2007) Nature 446:395–403
35. Arcadi A (2008) Chem Rev 108:3266–3325
36. Shapiro ND, Toste FD (2010) Synlett 675–691
37. Zificsak CA, Mulder JA, Hsung RP, Rameshkumar C, Wei LL (2001) Tetrahedron 57:7575–7606
38. Dekorver KA, Li H, Lohse AG, Hayashi R, Lu Z, Zhang Y, Hsung RP (2010) Chem Rev 110:5064–5106
39. Wang X-N, Yeom H-S, Fang L-C, He S, Ma Z-X, Kedrowski BL, Hsung RP (2014) Acc Chem Res 47:560–578

40. Pan F, Shu C, Ye L-W (2016) Org Biomol Chem 14:9456–9465
41. Silva C, Faza ON, Luna MM (2019) Front Chem 7:1–22
42. Couty S, Meyer C, Cossy J (2006) Angew Chem Int Ed 45:6726–6730
43. Zheng N, Chang Y-Y, Zhang L-J, Gong J-X, Yang Z (2016) Chem Asian J 11:371–375
44. Asiri AM, Hashmi ASK (2016) Chem Soc Rev 45:4471–4503
45. Schmidt B, Audörsch S (2016) Org Lett 18:1162–1165
46. Kramer S, Madsen JLH, Rottländer M, Skrydstrup T (2010) Org Lett 12:2758–2761
47. Sharp PP, Banwell MG, Renner J, Lohmann K, Willis AC (2013) Org Lett 15:2616–2619
48. Matsuda Y, Naoe S, Oishi S, Fujii N, Ohno H (2015) Chem Eur J 21:1463–1467
49. Hirano K, Inaba Y, Watanabe T, Oishi S, Fujii N, Ohno H (2010) Adv Synth Catal 352:368–372
50. Irikura KK, Goddard WA (1994) J Am Chem Soc 116:8733–8740
51. Raubenheimer HG, Esterhuysen MW, Timoshkin A, Chen Y, Frenking G (2002) Organometallics 21:3173–3181
52. Nieto-Oberhuber C, Muñoz MP, Buñuel E, Nevada C, Cárdenas DJ, Echavarren AM (2004) Angew Chem Int Ed 43:2402–2406
53. Mamane V, Gress T, Krause H, Fürstner A (2004) J Am Chem Soc 126:8654–8655
54. Luzung MR, Markham JP, Toste FD (2004) J Am Chem Soc 126:10858–10859
55. Gorin DJ, Davis NR, Toste FD (2005) J Am Chem Soc 127:11260–11261
56. Shapiro ND, Toste FD (2007) J Am Chem Soc 129:4160–4161
57. Tokimizu Y, Oishi S, Fujii N, Ohno H (2014) Org Lett 16:3138–3141

Chapter 2
Total Synthesis of Dictyodendrins by the Gold-Catalyzed Cascade Cyclization of Conjugated Diynes with Pyrroles

Abstract Total synthesis of dictyodendrins A–E was achieved on the basis of a novel gold-catalyzed cascade reaction for the construction of pyrrolo[2,3-c]carbazole scaffold. The synthetic strategy features functionalization of pyrrole pyrrolo[2,3-c]carbazole scaffold at the C1 (arylation), C2 (acylation), N3 (alkylation), and C5 (oxidation) positions. This synthetic method could be used for the diversity-oriented synthesis of dictyodendrin derivatives for medicinal applications.

As described in Chap. 1, dictyodendrins belong to a family of marine indole alkaloids having important bioactivities. The Fusetani group reported that dictyodendrins A–E have inhibitory activity against telomerase, thus making them potential lead as anticancer agent. Recent reports by Ready has shown that dictyodendrins F, H, and I displayed cytotoxicity against several human cancer cell lines [1]. Since dictyodendrins F and H–J exhibit inhibitory activity toward β-site amyloid-cleaving enzyme 1 (BACE1), they are also recognized as potential lead compounds for the treatment of Alzheimer's disease [2].

The most of the reported syntheses utilized the common strategy based on the introduction of requisite substituents to the cyclization precursors prior to the construction of the pyrrolo[2,3-c]carbazole core. The author envisaged that the development of a diversity-oriented synthesis of dictyodendrins on the basis of the early-stage construction of the core structure followed by regioselective introduction of the substituents would further accelerate their medicinal applications.

Homogeneous gold catalyst is recognized as effective catalysts for the electrophilic activation of alkynes [3–8]. Gold carbenoids have been emerged as the versatile intermediates for construction of polycyclic compounds in many gold-catalyzed transformations. Recently, Gagosz [9] and Zhang [10] reported the pioneering work on gold(I)-catalyzed intramolecular acetylenic Schmidt reactions using ethynylbenzene **A** (Scheme 2.1). In this reaction, gold carbenoid species **B** is generated by the gold-mediated nucleophilic attack of the azide moiety on the activated alkyne, followed by the elimination of nitrogen. Subsequent nucleophilic trapping of gold carbenoid species **C** afforded an indole **D** bearing an electron-donating substituent at the C3 position. However, to the best of the author's knowledge, there was no report

© The Editor(s) (if applicable) and The Author(s), under exclusive license to Springer Nature Singapore Pte Ltd. 2020
J. Matsuoka, *Total Synthesis of Indole Alkaloids*, Springer Theses, https://doi.org/10.1007/978-981-15-8652-1_2

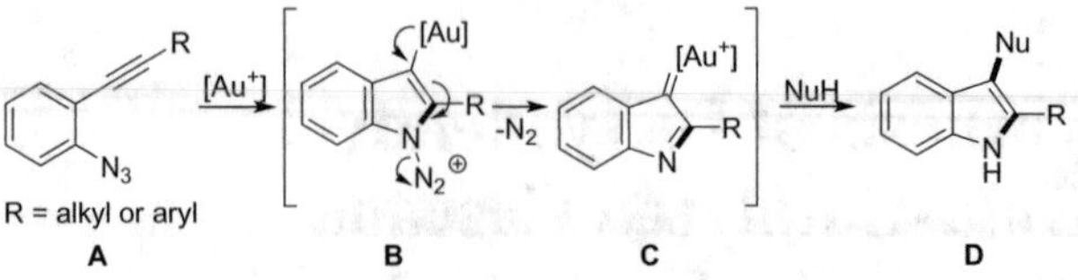

Scheme 2.1 Gold(I)-catalyzed intramolecular acetylenic Schmidt reaction

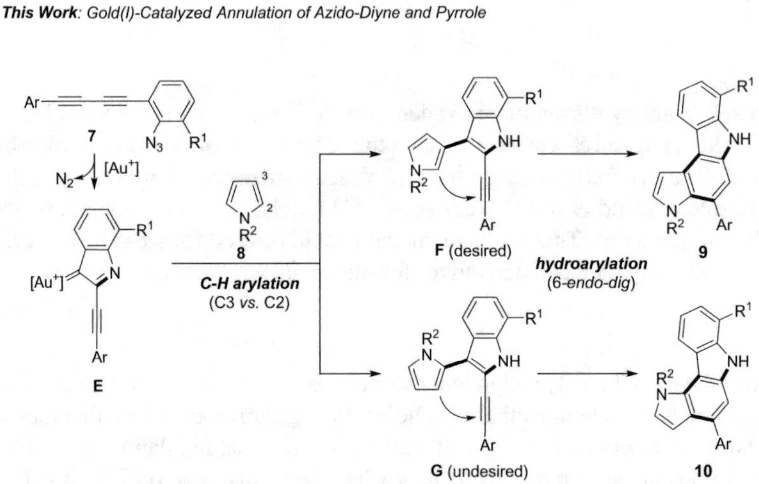

Scheme 2.2 Author's strategy for the construction of the pyrrolo[2,3-c]carbazole core structure

in the literature concerning to the reactivity of conjugated dynes with azides, in spite of the potential strategy of efficient construction for polycyclic compounds [11].

The author's research group have been engaged in development of gold-catalyzed cascade reactions for the synthesis of indole derivatives using various types of alkynes [12–16]. With the usefulness of the gold catalysis in acetylenic Schmidt reaction revealed, the author decided to develop a gold-catalyzed annulation of azido-diyne 7 with pyrrole 8 for the synthesis of pyrrolo[2,3-c]carbazole 9 (Scheme 2.2). Thus, the gold-catalyzed reaction of 7 would lead to formation of alkyne-substituted gold carbenoid E via nucleophilic attack of the azido group followed by elimination of nitrogen. Subsequent arylation of the carbenoid E with pyrrole 8 at the C3 position would give the pyrrole-substituted 2-alkynyl indole intermediate F, which would then undergo 6-*endo-dig* intramolecular hydroarylation to afford the pyrrolo[2,3-c]carbazole derivative 9 having the dictyodendrin core structure. On the other hand, the arylation of the gold carbenoid E with the pyrrole 2-position would produce the regioisomeric pyrrolo[3,2-c]carbazole 10. The author expected that the regioselectivity in the first arylation step could be controlled by appropriate tuning of the catalyst or substrate structure to obtain the desired pyrrolocarbazole 9 [17]. In this chapter, the author describes diversity-oriented total and formal syntheses of dictyodendrins A–F

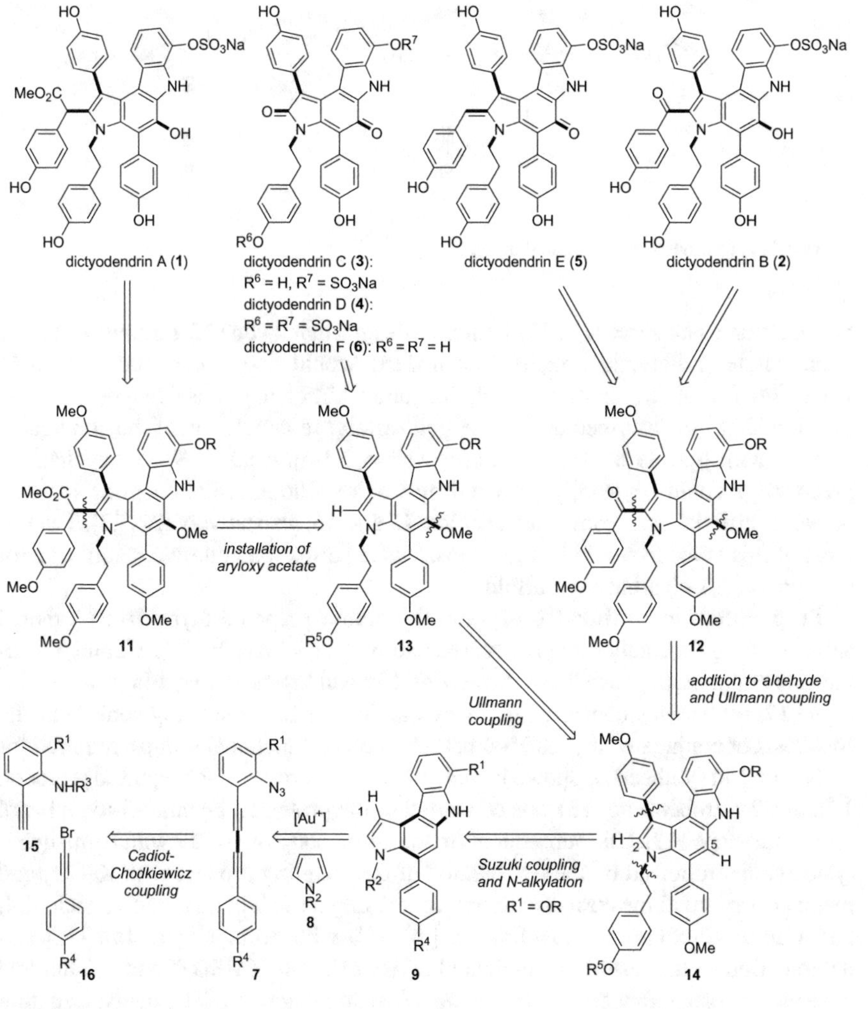

Scheme 2.3 Retrosynthetic analysis of dictyodendrins

on the basis of the gold-catalyzed direct annulation of the pyrrolo[2,3-c]carbazole core [18]. Biological evaluations of their derivatives are also presented.

The author's retrosynthetic analysis of dictyodendrins based on the designed gold-catalyzed annulation is shown in Scheme 2.3. Dictyodendrin A (1) can be synthesized from 11 by installation of a sulfate group and removal of methyl groups, according to Tokuyama's protocol [19]. The ester 11 can be prepared from 13 through a Friedel–Crafts reaction using a methoxyphenyl acetate. The pyrrolocarbazole 13, known as the precursor of dictyodendrins C (3), D (4), and F (6), could be obtained by bromination of 14 followed by Ullmann coupling with NaOMe [20]. The compound 14 would be prepared by the consecutive functionalization of 9 at the C1 and N3 positions, after

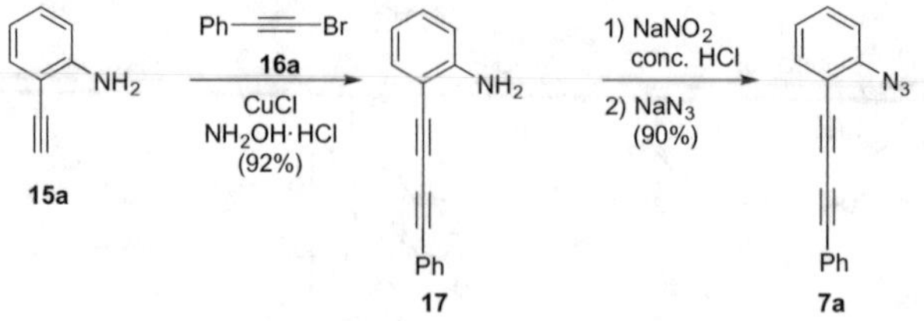

Scheme 2.4 Synthesis of conjugated diyne **7a**

bromination when necessary. The author envisaged that ketone **12**, a known synthetic intermediate of dictyodendrins B (**2**) and E (**5**), would also be constructed from **14** via a sequence of bromination, metalation, and addition to *p*-anisaldehyde at the C2 position [21]. As described above, the gold-catalyzed annulation of the conjugated diyne **7** with pyrrole **8** would produce pyrrolo[2,3-*c*]carbazole **9**. The cyclization precursor **7** could be readily prepared by Cadiot–Chodkiewicz coupling reaction between terminal and brominated alkynes, **15** and **16**, respectively [22, 23]. The key issue of this strategy would be the regioselective formation and functionalization of the pyrrolo[2,3-*c*]carbazole scaffold.

Preparation of conjugated diynes: The author prepared diyne **7a** as a model substrate for gold-catalyzed pyrrolocarbazole synthesis (Scheme 2.4). Cadiot–Chodkiewicz coupling between 2-ethynylaniline **15a** and bromoalkyne **16a** gave aminodiyne **17**, which was converted to **7a** by Sandmeyer reaction using sodium azide. Synthesis of conjugated diynes **7b-d** bearing oxygen functional groups, required for dictyodendrin synthesis, is shown in Scheme 2.5. According to the reported protocol, 1-fluoro-2-nitrobenzene (**18**) was converted to the protected 2-amino-3-iodophenol **19** in four steps [19]. The subsequent Sonogashira coupling of **19** with trimethylsilylacetylene, followed by the desilylation of the coupling product with K_2CO_3 and methanol, afforded the corresponding terminal alkyne **15b** in quantitative yield [24, 25]. Cadiot–Chodkiewicz coupling [22] of **15b** with bromoalkyne **16b** [26] and deprotection of the resulting conjugated diyne **21b** with TMSOTf and 2,6-lutidine gave the corresponding diyne **22b** having a free amino group [27]. Finally, azidation of **22b** using *t*-BuONO and $TMSN_3$ afforded the substrate **7b** with a *t*-Bu protection [28]. Other conjugated diynes **7c** (R = Ms) and **7d** (R = Bn) were prepared from **21b** in the similar manner through protecting group modifications and azidation as shown in Scheme 2.5.

Gold-catalyzed annulation of conjugated diynes with pyrrole: The author then explored the optimal conditions for the gold-catalyzed annulation using the model substrate **7a** (Table 2.1). Treatment of **7a** and unprotected pyrrole **8a** with BrettPhosAuSbF$_6$ (5 mol%) in dichloroethane (DCE) at 80 °C led to complete consumption of the starting material, leading to formation of an isomeric mixture of the two annulation products **9aa** and **10aa** in *ca.* 62% yield along with several impurities. Spectroscopic analyses of the isolated regioisomers revealed that, unfortunately, the undesired isomer **10aa** was obtained as the major product (**9/10** = 25:75; Table 2.1, entry 1). Considering that this would be the result of the higher

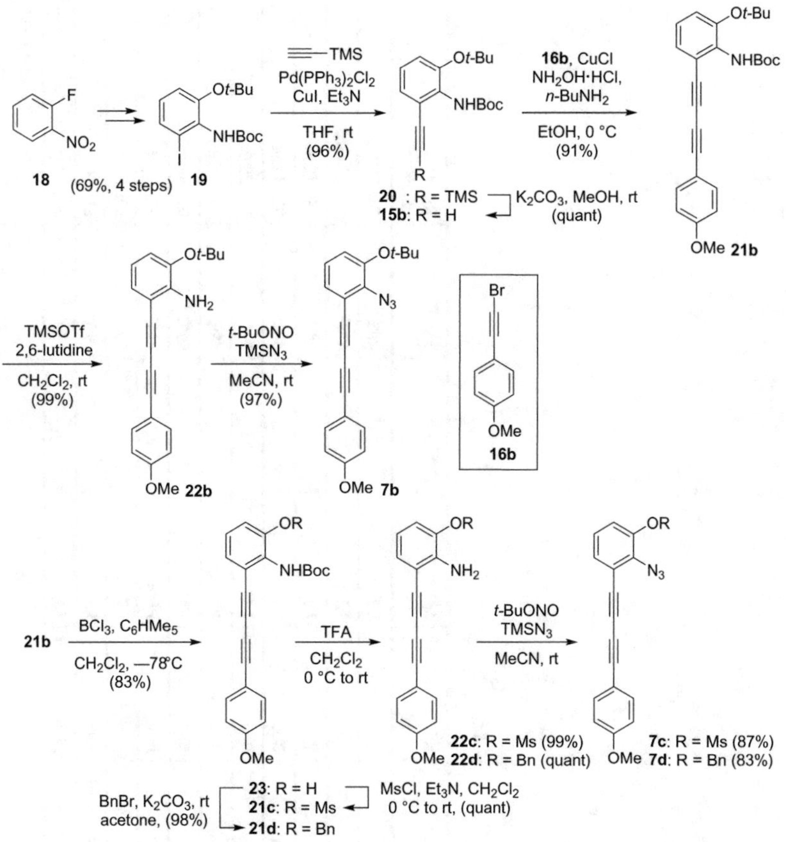

Scheme 2.5 Synthesis of azido-diynes **7b-d**

nucleophilicity of the C2 position of pyrrole **8a** relative to the C3 position, the author next evaluated the impact of different substituents at the pyrrole nitrogen (Table 2.1, entry 2–4). Rewardingly, by using *N*-Boc-pyrrole **8d**, the arylation and cyclization sequence proceeded with the required regioselectivity to afford the desired pyrrolo[2,3-*c*]carbazole **9ad** (**9/10** = 92:8; entry 4). Whereas the reaction at the room temperature decreased the regioselectivity for **9ad** (entry 5), the reaction at 140 °C in 1,1,2,2-tetrachloroethane (TCE) slightly improved the selectivity (**9/10** = 95:5; entry 6) [10]. Several other ligands including IPr, PPh$_3$, and JohnPhos (**L2**) were found to be much less effective than BrettPhos (**L1**) in terms of the regioselectivity and product yields (entries 7–9).

Based on the successful model reaction, the author focused on the preparation of the target pyrrolo[2,3-*c*]carbazoles bearing oxygen functional groups. The reactions of **7b** bearing methoxy and *tert*-butoxy groups (R^2 = O*t*-Bu) showed a slightly decreased regioselectivity (**9/10** = 84:16) with a good combined yield (79%). On the other hand, **7c** (R^2 = OMs) and **7d** (R^2 = OBn) gave the annulation products with

Table 2.1 Optimization of gold-catalyzed annulation of 1,3-diyne and pyrrole.[a]

Entry	Diyne	Pyrrole	R^1	R^2	R^3	L^b	T (°C)	Yield[c]	Ratio[d] (9: 10)
1	7a	8a	H	H	H	L1		<62%	25: 75 (aa)
2	7a	8b	H	H	Bn	L1		62%	18: 82 (ab)
3	7a	8c	H	H	Ts	L1		34%	58: 42 (ac)
4	7a	8d	H	H	Boc	L1		60%	92: 8 (ad)
5	7a	8d	H	H	Boc	L1		59%	72: 28 (ad)
6[e]	7a	8d	H	H	Boc	L1		58%	95: 5 (ad)
7[e]	7a	8d	H	H	Boc	IPr		60%	91: 9 (ad)
8[e]	7a	8d	H	H	Boc	PPh₃		<5%	87: 13 (ad)
9[e]	7a	8d	H	H	Boc	L2		56	92: 8 (ad)
10	7b	8d	OMe	Ot-Bu	Boc	L1		79%	84: 16 (bd)
11	7c	8d	OMe	OMs	Boc	L1		83%	75: 25 (cd)

(continued)

Table 2.1 (continued)

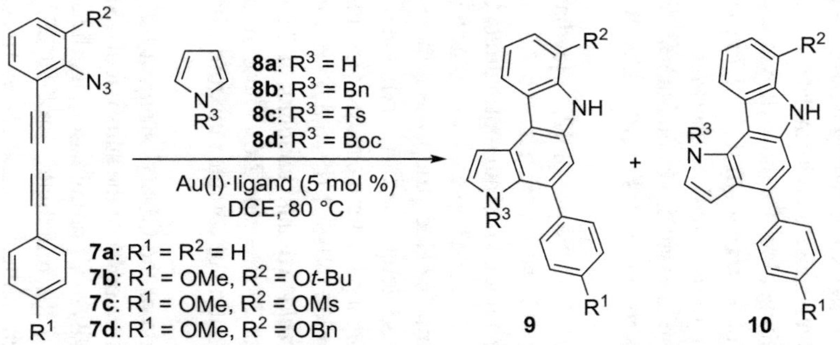

8a: R³ = H
8b: R³ = Bn
8c: R³ = Ts
8d: R³ = Boc

Au(I)·ligand (5 mol %)
DCE, 80 °C

7a: R¹ = R² = H
7b: R¹ = OMe, R² = Ot-Bu
7c: R¹ = OMe, R² = OMs
7d: R¹ = OMe, R² = OBn

9   10

Entry	Diyne	Pyrrole	R¹	R²	R³	L[b]	T (°C)	Yield[c]	Ratio[d] (9: 10)
12	7d	8d	OMe	OBn	Boc	L1		68%	81: 19 (**dd**)

[a]Reaction conditions: **7** (1 equiv), **8** (5 equiv), Au(I)·ligand (5 mol%), 1,2-dichloroethane (DCE), 80 °C
[b]The ligand structures are shown below. Unless otherwise noted, the catalysts were prepared in situ by mixing AuCl· ligand with AgNTf₂. For BrettPhos, the BrettPhosAu(MeCN)SbF₆ catalyst was prepared in advance
[c]Combined isolated yields
[d]Determined by ¹H NMR
[e]The reaction was carried out in 1,1,2,2-tetrachloroethane (TCE) at 140 °C

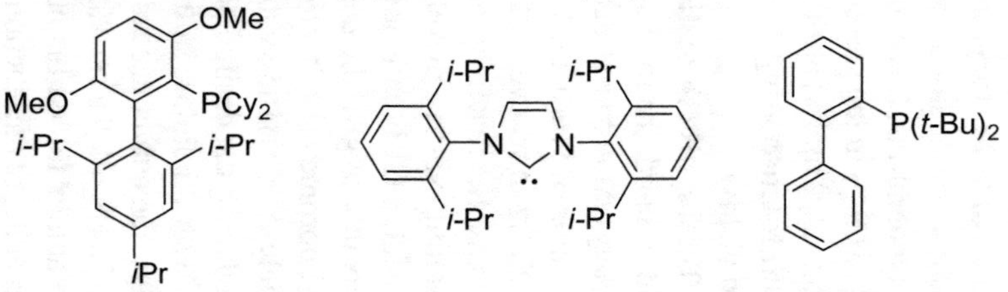

BrettPhos (**L1**)

IPr

JohnPhos (**L2**)

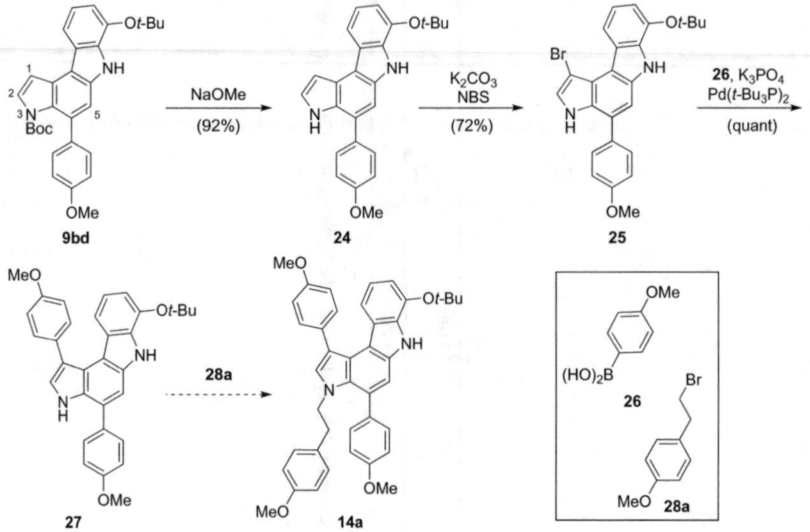

Scheme 2.6 Attempt at the synthesis of **14a**

relatively low regioselectivities (**9/10** = 75:25–81:19). Considering a facile depro-tection of *tert*-butyl group as well as a slightly better regioselectivity in the annulation reaction, the author decided **7b** as the suitable building block for the total synthesis of dictyodendrins. It is noteworthy that the reaction of **7b** on a gram scale (2.76 g) with **8d** (6.69 g) in the presence of a decreased amount of the BrettPhosAu(MeCN)SbF₆ (162 mg, 2 mol%) afforded **9bd** (2.27 g) in 58% isolated yield. The author's attempts at the reaction using 2- or 3-substituted pyrroles were unsuccessful (low selectivities and yields).

Total synthesis of dictyodendrins C, D, and F: With the pyrrolo[2,3-*c*]carbazole **9bd** in hand, the total synthesis of dictyodendrins C and F which have a 2,5-dioxo moiety on the core structure was investigated. The author's first attempt at direct C–H arylation of **9bd** at the C1 position with copper catalyst and hypervalent iodine failed, resulting in recovery of the starting material. Thus, the Boc group was removed to increase the reactivity of the pyrrole ring of **9bd**. Although the C–H arylation of the resulting N3-free pyrrolo[2,3-*c*]carbazole **24** using a palladium catalyst was unsuc-cessful, a C1 bromination with *N*-bromosuccinimide (NBS; 1.05 equiv) proceeded smoothly to give the desired product **25**. The Suzuki–Miyaura coupling of **25** with aryl boronic acid **26** afforded C1-arylated product **27** (Scheme 2.6). Quite unfortu-nately, *N*-alkylation with **28a** only led to the formation of complex mixture without producing the desired product **14a**.

With the failure in the strategy for introduction of the C1-aryl group at the first stage, the author needed to optimize the introduction order of the substituents. Since the first N-alkylation was found to decrease reactivity of C1 bromination significantly, the author turned his attention to N-alkylation of the brominated product **25**. The screening of the reaction conditions including electrophile, base, solvent, and reaction temperature has revealed that treatment of **25** with bromide **28a** and NaOH in THF afforded the desired N3-alkylated product **29** in 26% yield (Table 2.2, entry 5). Further optimization study has shown that addition of 18-crown-6 (3 equiv) using THF/H_2O (10: 1) as the reaction solvent improves the yield of **29** to 82% yield. It should be noted that gram-scale bromination for preparation of **25** was unsuccessful, presumably due to a rapid 'bromine dance' [21] during the evaporation of the reaction solvent. Thus, a one-pot C1-bromination/N3-alkylation protocol was employed for total synthesis of dictyodendrins.

The author next proceeded to formal and total synthesis of dictyodendrins C and F, respectively (Scheme 2.7). One-pot bromination of **24** with NBS (1.05 equiv) and N-alkylation with **28a** under the optimization conditions (Table 2.1, entry 7), followed by a Suzuki–Miyaura coupling with anisyl boronic acid (**26**), afforded **14a** having newly introduced substituents at the C1 and N3 positions. Introduction of an oxygen functional group at the C5 position was significantly difficult. After several unsuccessful attempts such as direct C-H borylation [29, 30] or lithiation [31, 32], the author finally succeeded in the formation of mono-bromide **31** through dibromination of **14a** with NBS (2.05 equiv) at the C2 and C5 positions followed by mono-selective debromination of **30** at C2 with $NaBH_4$ using a $PdCl_2(dppf)$ [33]. The Ullmann coupling of **31** with NaOMe in the presence of CuI gave **13a**, which is known as a precursor of dictyodendrin C as reported by Tokuyama [19]. The author completed the total synthesis of dictyodendrin F by deprotection of **13a** with BBr_3 and the subsequent aerobic oxidation [21].

Formal synthesis of dictyodendrin D (**4**) was achieved in a same manner (Scheme 2.8). Since dictyodendrin D has a sulfate group at the benzene ring of the N3 alkyl group, benzyl-protected bromide **28** was employed for alkylation following Tokuyama's synthesis [19]. Thus, the key intermediate **14b** was obtained from **24** through a sequence of reactions including C1-bromination, N3-alkylation with **28d**, and subsequent Suzuki–Miyaura coupling reaction. Formal total synthesis of dictyodendrin D was accomplished by introduction of a methoxy group into **14b**, providing the known precursor **13b** [19].

Total synthesis of dictyodendrins B and E: Dictyodendrins B and E possess acyl or benzylidene group at the C2 position, respectively (Scheme 2.9), which required an additional C–C bond formation for completion of their total synthesis. The author chose the C2 acylation strategy reported by Fürstner [21], where a sequence of reactions involving bromine–lithium exchange and addition of the resulting aryl lithium to *p*-anisaldehyde afforded the pyrrolo[2,3-*c*]carbazole carrying a benzyl alcohol moiety. Thus, a regioselective mono-bromination of **14a** was conducted with NBS (1.05 equiv) to give **32** in moderate yield (52%). The subsequent bromine–lithium exchange with MeLi (1.1 equiv) and *n*-BuLi (1.1 equiv) followed by addition to *p*-anisaldehyde afforded the corresponding C2-substituted product **33** in 74% yield. After selective mono-bromination of **33** at the C5 position, methyl ether **12a** was obtained by the Ley–Griffith oxidation of the resulting bromide **34** followed by the Ullmann coupling with NaOMe for introduction of a methoxy group. Tokuyama and co-workers reported that **12a** can be transformed to dictyodendrin E (**5**) by reduction of carbonyl group, demethylation, construction of the sulfate moiety, and oxidation

Table 2.2 Investigation of N-alkylation[a]

Entry	28 (equiv)	Base (equiv)	Solvent	Temp (°C)	Additive (equiv)	Yield (%)
1	28a(3)	K₂CO₃(10)	DMF	80		trace
2	28b (3)	Cs₂CO₃(10)	DMF	rt	–	0
3	28c(3)	K₂CO₃(10)	DMF	80	–	0
4	28a(1.2)	NaH (2.5)	DMF	80	–	0
5	28a(2)	NaOH (10)	THF	50	–	26
6	28a (10)	NaOH (15)	THF/H₂O (1: 1)	rt	18-C-6 (3)	0
7	28a (10)	NaOH (15)	THF/H₂O (10: 1)	rt	18-C-6 (3)	82

[a]Reaction conditions: substrate **25** (1 equiv), **28**, base (X equiv), solvent (0.05 M), additive (3 equiv where applicable)

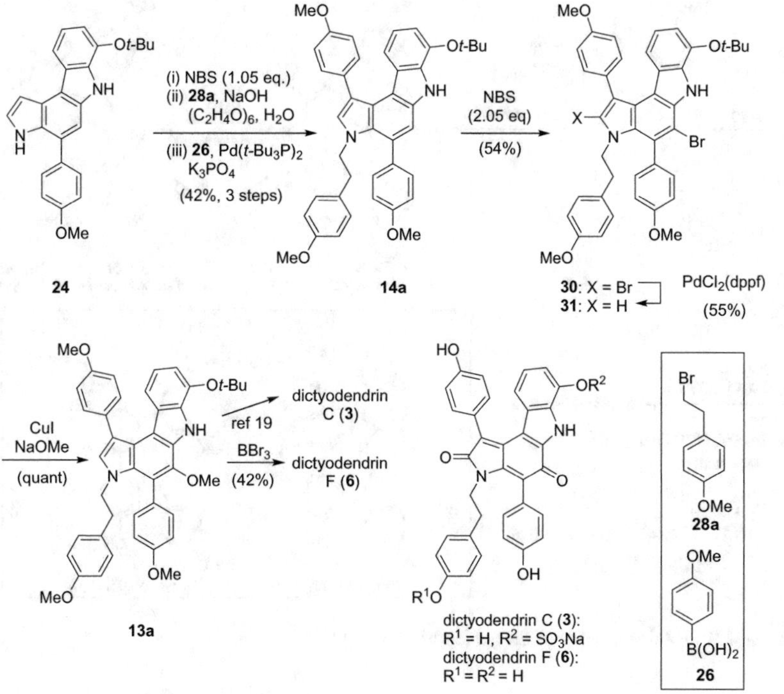

Scheme 2.7 Total syntheses of dictyodendrins C and F

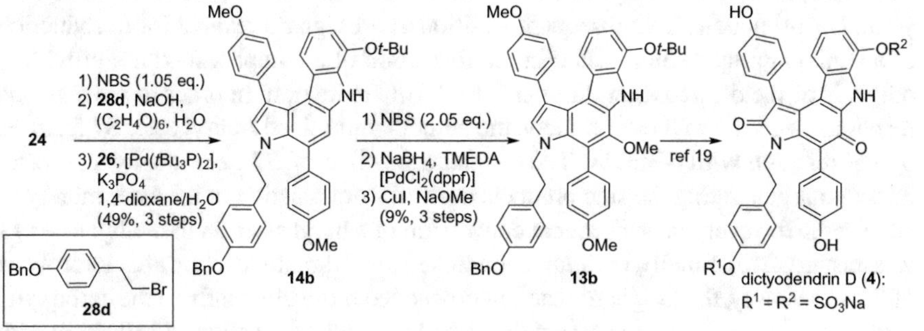

Scheme 2.8 Total synthesis of dictyodendrin D

with DDQ [19]. The author also accomplished the total synthesis of dictyodendrin B (**2**) by selective removal of *tert*-butyl group with BCl$_3$ at –78 °C, sulfate formation, and deprotection with BCl$_3$ (0 °C → rt) and Zn dust as reported [19].

Formal synthesis of dictyodendrin A: The author then focused on the total synthesis of dictyodendrin A (Scheme 2.10), which required the introduction of a (4-hydroxyphenyl)acetate moiety at the C2 position. The author's initial attempt at the introduction of a C2 substituent to **13a** including C-H insertion and Friedel–Crafts

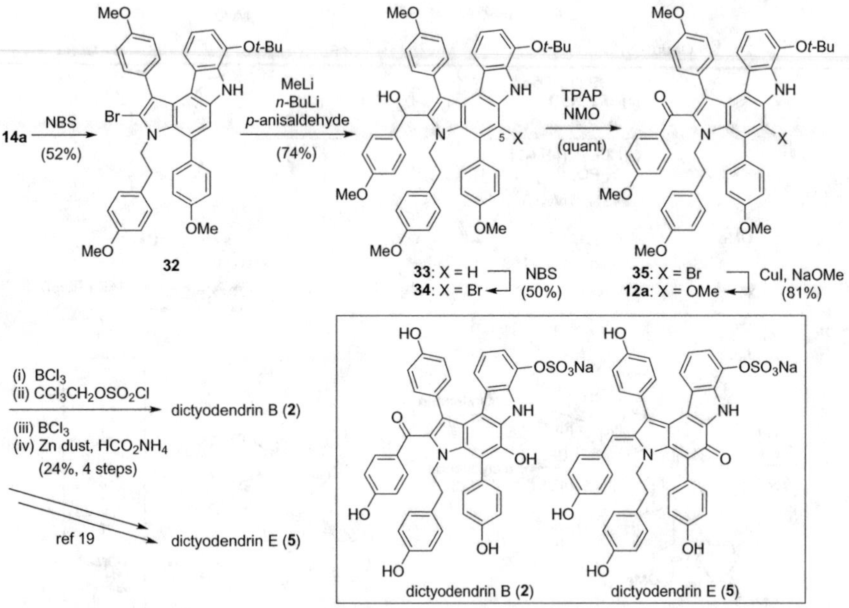

Scheme 2.9 Total syntheses of dictyodendrins B and E

reactions under several reaction conditions resulted in decomposition of the starting material. On the other hand, the author found that acylation of **13a** with oxalyl chloride followed by methyl esterification led to formation of keto-ester **36** in 87% yield. Unfortunately, the subsequent addition of a Grignard reagent for introduction of an anisyl group to **36** resulted in the formation of a complex mixture, providing only 9% of the desired ester **11** after Pd(OH)$_2$/C reduction. In order to prevent side reactions of the methyl ester moiety, the author synthesized t-butyl ester **37** from **36** by the reaction with t-BuOLi. The Grignard reaction of **37** gave a α-hydroxyester **38** without generating the side products derived from reaction of the ester moiety as expected. However, the subsequent conversion of t-butyl ester **38** to methyl ester **11** was unsuccessful. Finally, the author conducted the Grignard reaction after hydrolysis of **36**. Pleasingly, the Grignard reaction proceeded more efficiently to the carboxylic acid derived from **36**, giving rise to the ester **11** in 36% yield after esterification with TMS diazomethane and hydrogenation to remove the hydroxyl group. For a reason that is unclear, the ketone **12a** was observed during purification of hydroxy acid **39**. Thus, rapid esterification of **39** without purification was essential for the successful conversion. Finally, removal of the methyl and *tert*-butyl groups led to total synthesis of dictyodendrin A. Spectral data of all the synthetic natural products as well as the known intermediates were in good accordance with those reported the literatures [19].

   Biological evaluation: The resulting dictyodendrin analogues were applied to the development of novel bioactive substances. The author sought to evaluate biological

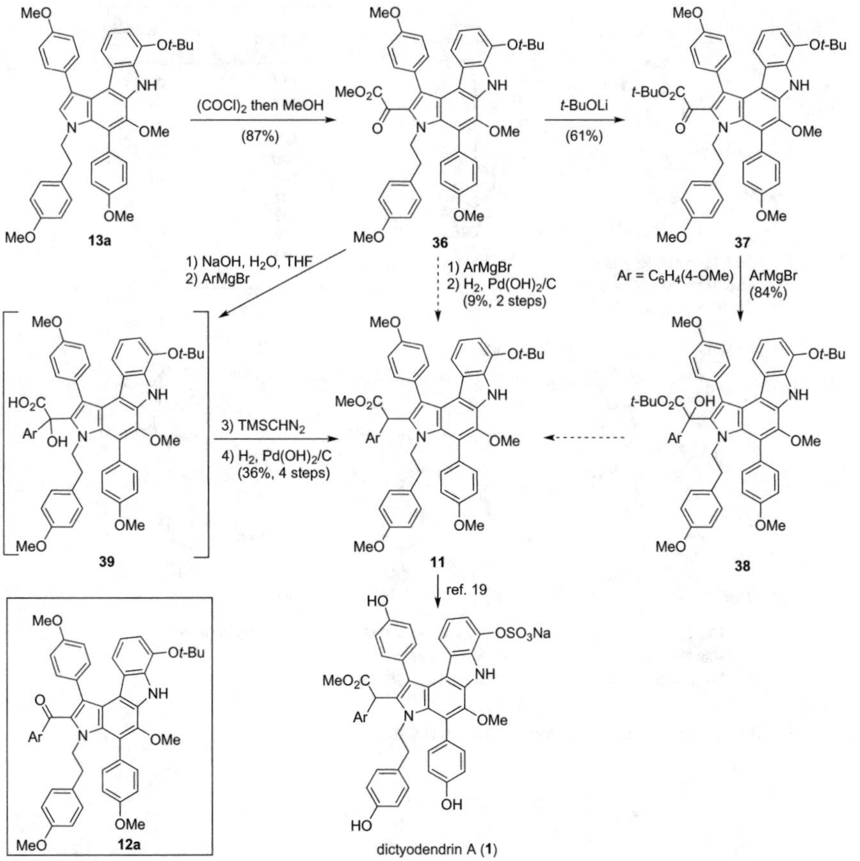

Scheme 2.10 Formal synthesis of dictyodendrin A

activities of dictyodendrin analogues, including the following: (1) cytotoxicity, (2) screening for kinase inhibition, and (3) inhibition of the nucleolar localization of Japanese encephalitis virus (JEV) core proteins.

(1) **Cytotoxicity evaluation**: The dictyodendrin F has been previously reported to show the cytotoxicity against human colon cancer HCT116 cells (IC$_{50}$ = 26.97 μM) [1]. For the purpose of identifying compounds displaying the antiproliferative activity, the author assessed the cytotoxicity of newly synthesized dictyodendrin analogues in HCT116 cells, using the colorimetric MTS assay (Fig. 2.1). Among the pyrrolo[2,3-c]carbazole derivatives investigated, the compound **24** without having substituents at the C1 and C2 positions exhibited relatively high cytotoxicity against HCT116 cell, comparable to dictyodendrin F. Interestingly, pyrrolo[3,2-c]carbazole derivative **10cd**, the unnatural

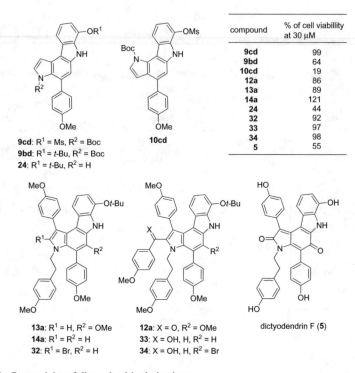

compound	% of cell viabillity at 30 μM
9cd	99
9bd	64
10cd	19
12a	86
13a	89
14a	121
24	44
32	92
33	97
34	98
5	55

9cd: R¹ = Ms, R² = Boc
9bd: R¹ = t-Bu, R² = Boc
24: R¹ = t-Bu, R² = H

10cd

13a: R¹ = H, R² = OMe
14a: R¹ = R² = H
32: R¹ = Br, R² = H

12a: X = O, R² = OMe
33: X = OH, H, R² = H
34: X = OH, H, R² = Br

dictyodendrin F (5)

Fig. 2.1 Cytotoxicity of dictyodendrin derivatives

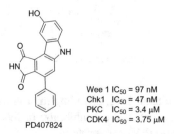

Wee 1 IC$_{50}$ = 97 nM
Chk1 IC$_{50}$ = 47 nM
PKC   IC$_{50}$ = 3.4 μM
CDK4 IC$_{50}$ = 3.75 μM

PD407824

Fig. 2.2 Inhibitory activity of PD407824 against protein kinases

regioisomer without having substitution at C1 and C2, showed the highest cyto-toxicity. In contrast, no inhibitory activity was observed with the corresponding substituted pyrrolocarbazoles such as **12a, 13a, 14a,** and **32–34** at 30 μM.

(2) **Screening for kinase inhibition**: Among the fused ring carbazoles, several compounds are known to exhibit inhibitory activities against protein kinases. For example, PD407824, containing a pyrrolo[3,4-c]carbazole scaffold, was reported to inhibit protein kinases (Wee1, Chk1, PKC, and CDK4) at the nM levels (Fig. 2.2). To examine the potential of pyrrolo[2,3-c]- and pyrrolo[3,2-c]carbazoles as the template for kinase inhibitors, the author then undertook a

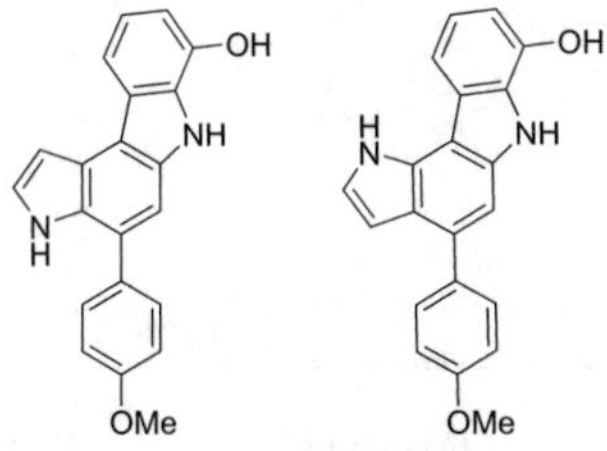

	IC$_{50}$ (μM)	
Compound	CDK2/CycA2	GSK3β
40	0.78	3.1
41	2.6	1.8

Fig. 2.3 Inhibition of CDK2/CycA2 and GSK3β

screening of unsubstituted derivatives **40** and **41** at 10 μM toward 32 protein kinases. As shown in Fig. 2.3, the author found that these compounds were potent inhibitors of CDK2/CycA2 [IC$_{50}$: 0.78 μM (**40**) and 2.6 μM (**41**)] and GSK3b [IC$_{50}$: 3.1 μM (**40**) and 1.8 μM (**41**)]. CDK2 belongs to the serine/threonine protein kinase family and involves in the progression of cells into the S- and M-phases of the cell cycle. In multiple cancer types, CDK2 activity is crucially associated with tumor growth, and thus CDK2 inhibitors have potential as anticancer agents. Glycogen synthase kinase 3β (GSK3β) is a multifunctional serine/threonine kinase that plays a critical role in regulating glycogen metabolism.

GSK3β also functions as a regulator of various biological processes, including cell cycle progression, proliferation, apoptosis signaling, and transcription. Thus, GSK3β has attracted much attention as a promising target for the treatment of diabetes and cancers.

(3) **Inhibition of the nucleolar localization of Japanese encephalitis virus (JEV) core protein**: Japanese encephalitis virus (JEV), which belongs to family of *Flaviviridae* including the genus *Flavivirus,* possesses a single-stranded positive-sense RNA as its genome. The life cycle of flaviviruses is previously thought to require only cytoplasm. However, Oka and Okamoto group demonstrated that the core protein of JEV was localized both in the cytoplasm and in the nucleus in host cells, and the mutant core protein displayed clearly disrupted nuclear localization. Furthermore, an in vivo study revealed that the nuclear localization of core protein was linked to the severity of JEV pathogenicity, suggesting that nuclear localization of the core protein is crucial for JEV propagation and pathogenicity. Recently, their group has developed a screening system to evaluate the inhibitory effects of compounds on the nuclear localization of the flavivirus core protein. As a result of evaluating various compounds, several CDK2 inhibitors were found to inhibit core protein nuclear translocation [34]. On the other hand, as mentioned above, the author revealed that dictyodendrin analogues **40** and **41** exhibited the inhibition against CDK2. Based on these findings, the author sought to evaluate the inhibitory activity of dictyodendrin analogues against the nuclear localization of the flavivirus core protein, using the developed screening system. Huh7 cell lines stably expressing the JEV core

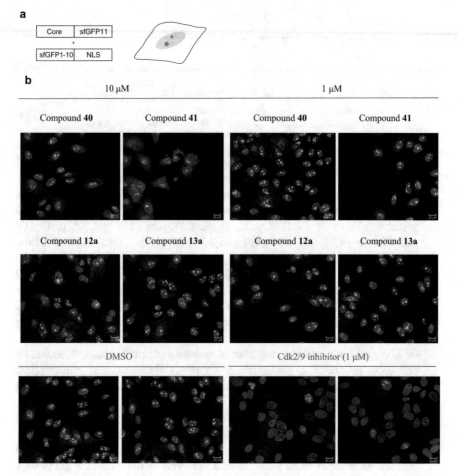

Fig. 2.4 **a** Schematic strategy for monitoring nuclear localization of the JEV core protein. Huh7 cells expressing the JEV core protein fused with split super-folder GFP 11 (sfGFP) and sfGFP1-10 fused with a non-classical nuclear localization signal (NLS) were generated and termed Huh7 sfGFP-JEV core. **b** Screening for nuclear localization of the JEV core protein. Huh7 sfGFP-JEV core was incubated with dictyodendrin analogues for 24 h. The cells were then fixed, and their GFP fluorescence was observed by confocal microscopy. DNA was stained with Hoechst 33342. Scale bars indicate 20 μm. DMSO-treated cells were used as a standard. Cdk2/9 inhibitor was used as a positive control

protein fused with split super-folder GFP 11 (sfGFP) and sfGFP1-10 fused with a non-classical nuclear localization signal (NLS) were used (Fig. 2.4a). As shown in Fig. 2.4b, fluorescence of the sfGFP-JEV core was detected in the nucleus of Huh7 cells when DMSO was used as a negative control. In contrast, the treatment with a known inhibitor (Cdk2/9 inhibitor) as a positive control resulted in greater than 50% reduction in fluorescence intensity compared to the intensity of DMSO treatment (Figs. 2.4b and 2.5). When the dictyodendrin

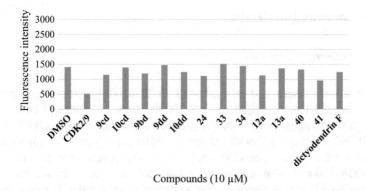

Compounds (10 μM)

Fig. 2.5 Huh7 sfGFP-JEV core was incubated with dictyodendrin analogues (10 μM in DMSO) for 24 h, and then the fluorescence intensity in the nucleus was observed using a CV7000S confocal microscope. The values plotted on the graph are the average fluorescence intensities. The value of the DMSO-treated cells was used as a standard. Cdk2/9 inhibitor was used as a positive control, which had 50% lower fluorescence intensities than DMSO

analogues were treated, some compounds displayed a slight decrease in fluorescence intensity, while no compounds were found to exhibit equal or higher activity than the known inhibitor (Cdk2/9 inhibitor).

Taken together, the author found that the pyrrolo[3,2-*c*]carbazole derivative **10cd** showed high cytotoxicity against HCT116 cells, and the pyrrolo[2,3-*c*]carbazole derivative **40** and its [3,2-*c*] congener **41** are a promising template for kinase inhibitors.

In conclusion, the author has accomplished the total and formal synthesis of dictyodendrin A (formal), B, C (formal), D (formal), E (formal), and F. A key discovery is that the regioselectivity of gold(I)-catalyzed annulation can be switched by substituents of pyrrole nitrogen atom, which allowed for efficient construction of the pyrrolo[2,3-*c*]carbazole scaffolds. The strategy of subsequent functionalization of the resulting pyrrolo[2,3-*c*]carbazole **9bd** at the C1 arylation, C2 acylation and bromination, N3 alkylation, and C5 oxidation served as diversity-oriented total synthesis of dictyodendrins. In addition, the resulting dictyodendrin analogues **40** and **41** exhibited potent for inhibition of CDK2/CycA2 and GSK3β. This strategy will enable the syntheses of related dictyodendrin derivatives, thus providing a new approach for the biologically interesting pyrrolocarbazole-type compounds.

2.1 Experimental Section

2.1.1 General Methods

IR spectra were determined on a JASCO FT/IR-4100 spectrometer. Exact mass (HRMS) spectra were recorded on JMS-700 mass spectrometer or Shimadzu LC-ESI-IT-TOF-MS equipment. ^{1}H NMR spectra were recorded using a JEOL AL-400 or JEOL ECA-500 or ECZ-600R. Chemical shifts are reported in δ (ppm) relative to Me_4Si (in $CDCl_3$) as internal standard. ^{13}C NMR spectra were recorded using a JEOL ECA-500 and referenced to the residual solvent signal. Melting points were measured by a hot stage melting point apparatus (uncorrected). For column chromatography, silica gel (Wakogel C-200E: Wako Pure Chemical Industries, Ltd.), amine silica gel (CHROMATOREX NH-DM1020: Fuji Silysia Chemical Ltd.), and diol silica gel (CHROMATOREX DIOL MB100-75/200: Fuji Silysia Chemical Ltd.) were employed. Purification by reverse-phase chromatography was performed using Cosmosil 5C18-ARII column (20.0 × 250 mm, Nacalai Tesque Inc.) with using acetonitrile/0.1% (v/v) TFA aq. as an eluent.

2.1.2 Preparation of the Cyclization Precursors 7

2-Ethynylaniline (15a). A mixture of 2-iodoaniline (4.38 g, 20.0 mmol), TMS acetylene (3.30 mL, 24.0 mmol), $PdCl_2(PPh_3)_2$ (284 mg, 0.405 mmol), CuI (76.0 mg, 0.400 mmol), and Et_3N (22.3 mL, 160 mmol) in THF (30 mL) was stirred at room temperature under Ar for 1.5 h. The mixture was filtered through a pad of Celite and concentrated in vacuo. The residue was purified by column chromatography on silica gel (hexane/EtOAc = 5/1) to give 2-[(trimethylsilyl)ethynyl]aniline [35] (3.65 g, 96%) as an orange oil. To a solution of this oil (1.69 g, 8.93 mmol) in MeOH (18.0 mL) was added K_2CO_3 (2.49 g, 18.8 mmol). The reaction mixture was stirred at room temperature for 20 min and concentrated in vacuo. The residue was diluted with Et_2O. The organic layer was washed with H_2O and brine, dried over $MgSO_4$, filtered, and concentrated in vacuo to give **15a** (994 mg, 94%) as a brown oil [36]. The spectral data were in good agreement with those previously reported [36].

    (Bromoethynyl)benzene(16a). To a solution of ethynylbenzene(1.87 mL, 17.0 mmol) in acetone (85 mL) were added NBS (3.33 g, 18.7 mmol) and $AgNO_3$ (289 mg, 1.70 mmol). The mixture was stirred at room temperature for 10 h. The mixture was diluted with n-hexane and filtered through a pad of silica gel. The filtrate was concentrated in vacuo to give **16a** (2.83 g, 92%) as a brown oil. The spectral data were in good agreement with those previously reported [37].

    2-[(4-Methoxyphenyl)buta-1,3-diyn-1-yl]aniline (17) [38]. To a mixture of **15a** (1.06 g, 9.05 mmol), CuCl (45.0 mg, 0.453 mmol), $NH_2OH \cdot HCl$ (253 mg, 3.64 mmol), and n-$BuNH_2$ (2.30 mL, 22.8 mmol) in dry EtOH (23 mL) was added a solution of **16a** (2.14 g, 11.8 mmol) in dry EtOH (5.0 mL) via dropping funnel at

0 °C under Ar. The mixture was stirred at room temperature for 1.5 h and concentrated in vacuo. The residue was diluted with Et$_2$O. The organic layer was washed with saturated aqueous NH$_4$Cl, H$_2$O, and brine, dried over Na$_2$SO$_4$, filtered, and concentrated in vacuo. The filtrate was purified by column chromatography on silica gel (hexane/EtOAc = 10/1) to give **17** (1.80 g, 92%) as a light yellow powder. The spectral data were in good agreement with those previously reported [26]: ^{1}H NMR (500 MHz, CDCl$_3$) δ: 4.32 (s, 2H), 6.69–6.70 (m, 2H), 7.16 (dd, J = 7.5, 1.5 Hz, 1H), 7.32–7.39 (m, 4H), 7.52–7.54 (m, 2H); ^{13}C NMR (125 MHz, CDCl$_3$) δ: 73.9, 78.7, 78.9, 82.6, 105.7, 114.2, 117.7, 121.5, 128.3 (2C), 129.0, 130.5, 132.2 (2C), 132.8, 149.5.

1-Azido-2-[(4-methoxyphenyl)buta-1,3-diyn-1-yl]benzene (7a) [39]. A solution of **17** (1.09 g, 5.02 mmol) in THF/H$_2$O/conc. HCl (1/1/1, 10 mL) was cooled to 0 °C. To the solution was added NaNO$_2$ (690 mg, 10.0 mmol) in H$_2$O (10 mL) via dropping funnel at 0 °C. After the mixture was stirred at 0 °C for 15 min, NaN$_3$ in H$_2$O was slowly added to the mixture at 0 °C, and the mixture was stirred for 3 h. The reaction mixture was diluted with H$_2$O. The resulting mixture was extracted with EtOAc twice. The combined organic layer was washed with brine, dried over MgSO$_4$, filtered, and concentrated in vacuo. The residue was purified by column chromatography on silica gel (hexane/EtOAc = 20/1) to give **7a** (1.09 g, 90%) as an light yellow powder: mp 93–96 °C; IR (neat) 2130 (C≡C), 2100 (C≡C); ^{1}H NMR (500 MHz, CDCl$_3$) δ: 7.11 (t, J = 7.0 Hz, 1H), 7.15 (d, J = 8.5 Hz, 1H), 7.33–7.41 (m, 4H), 7.50–7.54 (m, 3H); ^{13}C NMR (125 MHz, CDCl$_3$) δ: 73.6, 77.0, 79.3, 83.3, 113.8, 118.5, 121.6, 124.6, 128.4 (2C), 129.3, 130.4 132.5 (2C), 134.4, 142.6; HRMS (ESI$^+$) calcd for C$_{16}$H$_{10}$N$_3$ (MH$^+$): 244.0869, found 244.0867.

tert-Butyl [2-(_tert_-butoxy)phenyl]carbamate (S2) [19]. A solution of **18** (11.2 mL, 106 mmol) in THF (500 mL) was cooled to 0 °C. To the solution was added KOt-Bu (16.6 g, 148 mmol) in THF (100 mL) via dropping funnel at 0 °C. The mixture was warmed to room temperature and stirred for 1 h. The mixture was diluted with saturated aqueous NH$_4$Cl. The resulting mixture was extracted with CH$_2$Cl$_2$ twice. The combined organic layer was dried over Na$_2$SO$_4$ and filtered. The filtrate was concentrated in vacuo to give crude **S1** (21.1 g) as an orange oil, which was used to the next without further purification. A mixture of crude **S1**(21.1 g), 10% Pd/C (11.3 g, 10.6 mmol) in EtOAc (50 mL) and EtOH (50 mL), was stirred at room temperature under H$_2$ for 38 h. The mixture was filtered through a pad of Celite. The filtrate was concentrated in vacuo to give the corresponding amine (17.9 g) as a brown oil. A mixture of this crude amine (9.3 g) and Boc$_2$O (16.5 g, 75.5 mmol) was stirred at 100 °C for 1 h. The mixture was purified by column chromatography on silica gel (hexane/EtOAc = 15/1) to give **S2** (16.0 g, quant, 3 steps from **18**) as an off-white solid. The spectral data were in good agreement with those previously reported [19].

tert-Butyl [2-(_tert_-butoxy)-6-iodophenyl]carbamate (19) [19]. To a solution of **S2** (10.6 g, 40.0 mmol) in dry Et_2O (50 mL) was added t-BuLi (1.9 M in pentane, 50.0 mL) slowly at -20 °C under Ar. The mixture was stirred at -20 °C for 3 h and cooled to -78 °C. To the mixture was added a solution of I_2 (15.2 g, 60.0 mmol) in Et_2O (120 mL) via cannula at -78 °C. The mixture was warmed to room temperature and stirred under Ar for 1.5 h. The mixture was quenched with saturated aqueous $Na_2S_2O_3$. The resulting mixture was extracted with Et_2O twice. The combined organic layers were washed with saturated aqueous $Na_2S_2O_3$ and brine, dried over Na_2SO_4, filtered, and concentrated in vacuo. The residue was purified by column chromatography on silica gel (hexane/EtOAc = 10/1) to give **19** (10.7 g, 69%) as an off-white solid. The spectral data were in good agreement with those previously reported [19].

tert-Butyl {2-(_tert_-butoxy)-6-[(trimethylsilyl)ethynyl]phenyl}carbamate (20). A mixture of **19** (19.6 g, 50.0 mmol), TMS acetylene (8.30 mL, 60.0 mmol), $PdCl_2(PPh_3)_2$ (702 mg, 1.00 mmol), CuI (190 mg, 1.00 mmol), and Et_3N (34.8 mL, 250 mmol) in THF (100 mL) was stirred at room temperature under Ar for 2 h. The mixture was filtered through a pad of Celite and concentrated in vacuo. The residue was purified by column chromatography on silica gel (hexane/EtOAc = 10/1) to give **20** (17.3 g, 96%) as a pale brown solid: mp 74 °C; IR (neat) 3310 (NH), 2157 (C≡C), 1727 (C=O); H NMR (500 MHz, CDCl₃) δ: 0.24 (s, 9H), 1.35 (s, 9H), 1.50 (s, 9H), 6.31 (br s, 1H), 7.00–7.05 (m, 2H), 7.20–7.21 (m, 1H); ¹³C NMR (125 MHz, CDCl₃) δ: 0.00 (3C), 28.3 (3C), 28.9 (3C), 80.0, 80.4, 98.8, 102.0, 121.3, 124.3, 125.4, 128.0, 133.7, 150.0, 152.8; HRMS (FAB) calcd for $C_{20}H_{32}NO_3Si$ (MH⁺): 362.2146, found 362.2151.

tert-Butyl [2-(_tert_-butoxy)-6-ethynylphenyl]carbamate (15b). To a solution of **20** (7.41 g, 20.5 mmol) in MeOH (100 mL) was added K_2CO_3 (5.67 g, 42.8 mmol). The mixture was stirred at room temperature for 30 min and concentrated in vacuo. The residue was diluted with Et_2O. The organic layer was washed with H_2O and brine, dried over Na_2SO_4, filtered, and concentrated in vacuo to give **15b** (5.90 g, quant) as a brown solid: mp 107 °C; IR (neat) 3286 (NH), 3228 (C≡CH), 1721 (C=O); ¹H NMR (500 MHz, CDCl₃) δ: 1.36 (s, 9H), 1.49 (s, 9H), 3.24 (s, 1H), 6.32 (br s, 1H), 7.03–7.08 (m, 2H), 7.23 (dd, $J = 6.9, 2.3$ Hz, 1H); ¹³C NMR (125 MHz, CDCl₃) δ: 28.2 (3C), 28.9 (3C), 80.3, 80.5, 80.9, 81.3, 120.6, 124.3, 125.5, 128.1,

134.0, 149.8, 152.9; HRMS (FAB) calcd for $C_{17}H_{24}NO_3$ (MH$^+$): 290.1751, found 290.1753.

1-(2,2-Dibromovinyl)-4-methoxybenzene (S4) [26]. To a solution of **S3** (7.30 mL, 60.0 mmol) and CBr$_4$ (29.9 g, 90.2 mmol) in CH$_2$Cl$_2$ (300 mL) was added PPh$_3$ (47.2 g, 180 mmol) in CH$_2$Cl$_2$ (300 mL) via dropping funnel at 0 °C. The mixture was stirred at 0 °C for 15 min and concentrated in vacuo. The residue was diluted with CHCl$_3$ and filtered through a pad of Celite. The filtrate was concentrated in vacuo. The residue was purified by column chromatography on silica gel (hexane/EtOAc = 20/1) to give **S4** (16.4 g, 94%) as a pale yellow solid. The spectral data were in good agreement with those previously reported [26].

1-(Bromoethynyl)-4-methoxybenzene (16b) [26]. To a solution of **S4** (19.3 g, 66.1 mmol) in CH$_2$Cl$_2$ (330 mL) were successively added BnEt$_3$NCl (13.2 g, 58.0 mmol) and a solution of KOH (98 g, 1.75 mol) in H$_2$O (130 mL) at 0 °C. The mixture was stirred at 0 °C for 4 h. To the mixture was added H$_2$O. The resulting mixture was extracted with CH$_2$Cl$_2$ twice. The combined organic layers were washed with brine, dried over MgSO$_4$, filtered, and concentrated in vacuo. The residue was purified by column chromatography on silica gel (hexane) to give **16b** (12.5 g, 89%) as a white solid. The spectral data were in good agreement with those previously reported [26].

tert-**Butyl** **{2-(*tert*-butoxy)-6-[(4-methoxyphenyl)buta-1,3-diyn-1-yl]phenyl}carbamate (21b)**. To a solution of **15b** (2.84 g, 9.81 mmol), CuCl (48.0 mg, 0.485 mmol), NH$_2$OH·HCl (270 mg, 3.89 mmol), and *n*-BuNH$_2$ (2.50 mL, 25.2 mmol) in dry EtOH (25 mL) was added a solution of **16b** (2.70 g, 12.8 mmol) in dry EtOH (17 mL) via dropping funnel at 0 °C over 30 min. The mixture was stirred at 0 °C for 3 h and concentrated in vacuo. The residue was diluted with EtOAc. The organic layer was washed with saturated aqueous NH_4Cl, H_2O, and brine, dried over Na_2SO_4, filtered, and concentrated in vacuo. The residue was precipitated with EtOAc/hexane to give **21b** as a pale yellow powder. The filtrate was purified by column chromatography on silica gel ($CHCl_3$/hexane = 1/1 to hexane/EtOAc = 2/1) and recrystallized from EtOAc/hexane to give **21b** (3.76 g) as a pale brown powder (91% total yield): mp 152 °C; IR (neat) 3493 (NH), 2213 (C≡C), 2143 (C≡C), 1698 (C=O); ^{1}H NMR (500 MHz, $CDCl_3$) δ: 1.36 (s, 9H), 1.53 (s, 9H), 3.82 (s, 3H), 6.35 (br s, 1H), 6.84–6.86 (m, 2H), 7.03–7.07 (m, 2H), 7.25 (dd, $J = 6.9, 2.3$ Hz, 1H), 7.43–7.45 (m, 2H); ^{13}C NMR (125 MHz, $CDCl_3$) δ: 28.2 (3C), 28.8 (3C), 55.3, 73.0, 78.1, 78.4, 80.5, 80.7, 82.4, 113.8, 114.1 (2C), 120.4, 124.6, 125.4, 128.4, 134.0 (2C), 134.6, 149.5, 152.9, 160.2; HRMS (FAB) calcd for $C_{26}H_{29}NO_4Na$ (MNa$^+$): 442.1989, found 442.1999.

2-(*tert*-Butoxy)-6-[(4-methoxyphenyl)buta-1,3-diyn-1-yl]aniline (22b) [28]. To a solution of **21b** (5.03 g, 12.0 mmol) and 2,6-lutidine (8.30 mL, 71.7 mmol) in dry CH_2Cl_2 was added TMSOTf (6.50 mL, 35.9 mmol) dropwise at 0 °C. The mixture was stirred at room temperature under Ar. After being stirred for 1 h, the mixture was treated with MeOH (24 mL) and stirred for 30 min, and H_2O (24 mL) was added. The resulting mixture was diluted with CH_2Cl_2 (120 mL) and stirred for 15 min. The organic layer was washed with H_2O three times and brine, dried over Na_2SO_4, filtered, and concentrated in vacuo. The residue was purified by column chromatography on silica gel ($CHCl_3$/hexane = 3/2–2/1) to give **22b** (3.78 g, 99%) as an off-white solid: mp 144 °C; IR (neat) 3375 (NH), 2210 (C≡C), 2135 (C≡C); ^{1}H-NMR ($CDCl_3$) δ: 1.40 (s, 9H), 3.82 (s, 3H), 4.45 (s, 2H), 6.54 (dd, $J = 8.0, 8.0$ Hz, 1H), 6.86 (d, $J = 8.6$ Hz, 2H), 6.95 (d, $J = 8.0$ Hz, 1H), 7.05–7.07 (m, 1H), 7.46 (d, $J = 8.6$ Hz, 2H); ^{13}C NMR (125 MHz, $CDCl_3$) δ: 28.9 (3C), 55.3, 72.8, 78.1, 79.1, 80.2, 82.8, 106.6, 113.8, 114.1 (2C), 116.7, 123.6, 127.3, 134.0 (2C), 142.2, 144.9, 160.3; HRMS (FAB) calcd for $C_{21}H_{22}NO_2$ (MH$^+$): 320.1645, found 320.1651.

2-Azido-1-(*tert*-butoxy)-3-[(4-methoxyphenyl)buta-1,3-diyn-1-yl]benzene (7b) [27]. To a solution of **22b** (3.67 g, 11.5 mmol) in MeCN (23 mL) were successively added *t*-BuONO (2.10 mL, 17.3 mmol) and TMSN$_3$ (1.80 mL, 13.8 mmol) at 0 °C. The mixture was stirred at room temperature for 1.5 h and concentrated in vacuo. The residue was purified by column chromatography (EtOAc/hexane = 1/10) to give **7b** (3.86 g, 97%) as a pale brown powder: mp 84–85 °C; IR (neat) 2130 (C≡C), 2087 (C≡C); ^{1}H-NMR ($CDCl_3$) δ: 1.43 (s, 9H), 3.82 (s, 3H), 6.84–6.87 (m, 2H), 6.97 (dd, $J = 8.0, 8.0$ Hz, 1H), 7.05–7.07 (m, 1H), 7.17–7.19 (m, 1H), 7.46–7.48 (m, 2H); ^{13}C NMR (125 MHz, $CDCl_3$) δ: 28.3 (3C), 55.3, 72.6, 77.2, 78.9, 81.7, 83.4, 113.6, 114.1 (2C), 115.6, 124.3, 124.4, 128.3, 134.1 (2C), 136.1, 149.5, 160.4. *Anal.*, calcd for $C_{21}H_{19}N_3O_2$: C, 73.03; H, 5.54; N, 12.17, found: C, 73.06; H, 5.66; N, 12.14.

tert-Butyl   **{2-hydroxy-6-[(4-methoxyphenyl)buta-1,3-diyn-1-yl]phenyl}carbamate (23)** [19]. To a solution of **21b** (1.05 g, 2.50 mmol) and C_6HMe_5 (1.11 g, 7.49 mmol) in CH_2Cl_2 (25 mL) was added BCl$_3$ (1 M in heptane, 6.30 mL) dropwise at –78 °C. The mixture was stirred at –78 °C for 15 min. The mixture was diluted with MeOH/CHCl$_3$ (1/10) at –78 °C, warmed to room temperature, and concentrated in vacuo. The residue was purified by column chromatography on silica gel (hexane/EtOAc = 5/1 to CHCl$_3$) and recrystallized from CHCl$_3$/hexane to give **23** (662 mg) as a white solid. The filtrate was concentrated in vacuo. The residue was purified by column chromatography on silica gel (CHCl$_3$/hexane = 3/1) to give **23** (91.7 mg) as a white solid (83% total yield): mp 174 °C; IR (neat) 3356 (NH), 2213 (C≡C), 2145 (C≡C), 1671 (C=O); ^{1}H NMR (500 MHz, CDCl$_3$) δ: 1.57 (s, 9H), 3.83 (s, 3H), 6.88 (d, J = 8.5 Hz, 2H), 7.02–7.04 (m, 2H), 7.09 (dd, J = 6.0, 3.0 Hz, 1H), 7.14 (s, 1H), 7.49–7.50 (m, 2H), 9.30 (s, 1H); ^{13}C NMR (125 MHz, CDCl$_3$) δ: 28.2 (3C), 55.4, 72.2, 76.4, 80.3, 83.3, 83.8, 113.2, 114.2 (2C), 114.4, 121.8, 125.4, 125.8, 127.4, 134.2 (2C), 148.4, 155.6, 160.6; HRMS (FAB) calcd for C$_{22}$H$_{21}$NNaO$_4$ (MNa$^+$): 386.1363, found 386.1360.

2-[(*tert*-Butoxycarbonyl)amino]-3-[(4-methoxyphenyl)buta-1,3-diyn-1-yl]phenyl methanesulfonate (21c). To a solution of **23** (180 mg, 0.495 mmol) and triethylamine (103 μL, 0.743 mmol) in CH$_2$Cl$_2$ (1.0 mL) was added mesyl chloride (68.0 mg, 0.594 mmol) dropwise at 0 °C. The mixture was stirred at room temperature for 10 min. The mixture was diluted with water. The resulting mixture was extracted with CH$_2$Cl$_2$ twice. The combined organic layers were washed with brine, dried over Na$_2$SO$_4$, filtered, and concentrated in vacuo. The residue was purified by column chromatography on silica gel (CHCl$_3$/hexane = 6/1) to give **21c** (218 mg, quant) as a white solid: mp 156 °C; IR (neat) 3317 (NH), 2216 (C≡C), 2130 (C≡C), 1713 (C=O); ^{1}H NMR (500 MHz, CDCl$_3$) δ: 1.53 (s, 9H), 3.20 (s, 3H), 3.83 (s, 3H), 6.55 (s, 1H), 6.87 (d, J = 8.6 Hz, 2H), 7.23 (dd, J = 8.0, 8.0 Hz, 1H), 7.38 (dd, J = 8.0, 1.1 Hz, 1H), 7.46 (d, J = 8.6 Hz, 2H), 7.49 (dd, J = 8.0, 1.1 Hz, 1H); ^{13}C NMR (125 MHz, CDCl$_3$) δ: 28.1 (3C), 37.9, 55.3, 72.4, 76.3, 80.0, 81.4, 83.7, 113.2, 114.2 (2C), 122.5, 124.1, 127.0, 132.3, 132.8, 134.1 (2C), 143.9, 152.7, 160.5; HRMS (FAB) calcd for C$_{23}$H$_{23}$NNaO$_6$S (MH$^+$): 464.1138, found 464.1144.

2-Amino-3-[(4-methoxyphenyl)buta-1,3-diyn-1-yl]phenyl methanesulfonate (22c). TFA (1.10 mL) was added dropwise to a solution of **21c** (552 mg, 1.25 mmol) in CH$_2$Cl$_2$ (6.3 mL) at 0 °C. The mixture was stirred at room temperature for 3 h. The mixture was diluted with 1 M NaOH and neutralized with 1 M HCl. The resulting mixture was extracted with CH$_2$Cl$_2$ twice. The combined organic layers were washed with brine, dried over Na$_2$SO$_4$, filtered, and concentrated in vacuo to give **22c** (424 mg, 99%) as a brown solid: mp 150 °C; IR (neat) 3449 (NH), 2215 (C≡C), 2140 (C≡C); ^{1}H NMR (500 MHz, CDCl$_3$) δ: 3.19 (s, 3H), 3.83 (s, 3H), 4.61 (s, 2H), 6.68 (dd, J = 8.0, 8.0 Hz, 1H), 6.86–6.89 (m, 2H), 7.24 (dd, J = 8.0, 1.5 Hz, 1H), 7.30 (dd, J = 7.5, 1.5 Hz, 1H), 7.46–7.49 (m, 2H); ^{13}C NMR (125 MHz, CDCl$_3$) δ: 37.8, 55.3, 72.3, 76.2, 80.5, 83.8, 109.3, 113.3, 114.2 (2C), 117.6, 124.1, 131.8, 134.1 (2C), 136.0, 142.6, 160.5; HRMS (ESI$^+$) calcd for C$_{18}$H$_{16}$NO$_4$S (MH$^+$): 342.0795, found 342.0793.

2-Azido-3-[(4-methoxyphenyl)buta-1,3-diyn-1-yl]phenyl methanesulfonate (7c) [27]. To a solution of **22c** (341 mg, 1.00 mmol) in MeCN (2.0 mL) were successively added *t*-BuONO (180 μL, 1.50 mmol) and TMSN$_3$ (0.160 μL, 1.20 mmol) dropwise at 0 °C. The mixture was stirred at room temperature under Ar for 15 min and concentrated in vacuo. The residue was purified by column chromatography on silica gel (CHCl$_3$/hexane = 4/1) to give **7c** (321 mg, 87%) as a yellow solid:

mp 133 °C; IR (neat) 3535 (NH), 2213 (C≡C), 2102 (C≡C); ^{1}H NMR (500 MHz, CDCl$_3$) δ: 3.29 (s, 3H), 3.84 (s, 3H), 6.87 (d, J = 8.6 Hz, 2H), 7.12 (dd, J = 8.0, 8.0 Hz, 1H), 7.34–7.38 (m, 1H), 7.43 (dd, J = 8.0, 1.1 Hz, 1H), 7.49 (d, J = 8.6 Hz, 2H); ^{13}C NMR (125 MHz, CDCl$_3$) δ: 38.3, 55.4, 72.1, 74.8, 82.0, 84.8, 113.1, 114.2 (2C), 117.8, 125.0, 125.4, 133.0, 134.3 (2C), 134.4, 141.4, 160.7; HRMS (FAB) calcd for C$_{18}$H$_{13}$N$_3$NaO$_4$S (MNa$^+$): 390.0519, found 390.0524.

tert-**Butyl**   **{2-(benzyloxy)-6-[(4-methoxyphenyl)buta-1,3-diyn-1-yl]phenyl}carbamate (21d)**. A mixture of **23** (545 mg, 1.50 mmol), BnBr (270 μL, 2.25 mmol), and K$_2$CO$_3$ (415 mg, 3.00 mmol) in dry acetone (3.0 mL) was stirred at room temperature for 19 h. To the mixture was added the second portion of BnBr (100 μL, 0.840 mmol). After being stirred for additional 5 h, the mixture was filtered through a pad of Celite. The filtrate was concentrated in vacuo. The residue was purified by column chromatography on silica gel (CHCl$_3$/hexane = 3/1) to give **21d** (668 mg, 98%) as a white solid: mp 144 °C; IR (neat) 3287 (NH), 2309 (C≡C), 2140 (C≡C), 1718 (C=O); ^{1}H NMR (500 MHz, CDCl$_3$) δ: 1.52 (s, 9H), 3.83 (s, 3H), 5.10 (s, 2H), 6.28 (br s, 1H), 6.86 (d, J = 9.2 Hz, 2H), 6.97 (d, J = 7.4 Hz, 1H), 7.09 (dd, J = 8.0, 8.0 Hz, 1H), 7.14–7.16 (m, 1H), 7.32–7.35 (m, 1H), 7.39 (dd, J = 7.4, 7.4 Hz, 2H), 7.42–7.46 (m, 4H); ^{13}C NMR (125 MHz, CDCl$_3$) δ: 28.2 (3C), 55.3, 70.7, 72.9, 78.0, 78.5, 80.5, 82.6, 113.72, 113.78, 114.1 (2C), 120.5, 125.7, 126.2, 127.4 (2C), 128.0, 128.5 (2C), 129.2, 134.0 (2C), 136.4, 152.8, 153.1, 160.3; HRMS (FAB) calcd for C$_{29}$H$_{27}$NO$_4$Na (MNa$^+$): 476.1832, found 476.1833.

2-(Benzyloxy)-6-[(4-methoxyphenyl)buta-1,3-diyn-1-yl]aniline (22d). To a solution of **21d** (181 mg, 0.399 mmol) in CH$_2$Cl$_2$ (2.0 mL) was added TFA (360μL) dropwise at 0 °C. The mixture was stirred at room temperature for 1 h. The mixture was diluted with 1 N NaOH aqueous solution and neutralized with 1 N HCl aqueous solution. The resulting mixture was extracted with CH$_2$Cl$_2$ twice. The combined organic layers were washed with brine, dried over Na$_2$SO$_4$, filtered, and concentrated in vacuo to give **22d** (145 mg, quant) as a brown solid: mp 134 °C; IR (neat) 3384 (NH), 2311 (C≡C), 2133 (C≡C); ^{1}H NMR (500 MHz, CDCl$_3$) δ: 3.80 (s, 3H), 4.50 (br s, 2H), 5.06 (s, 2H), 6.58 (dd, J = 8.0, 8.0 Hz, 1H), 6.81 (d, J = 8.0 Hz, 1H), 6.85 (d, J = 8.6 Hz, 2H), 6.98 (d, J = 8.0 Hz, 1H), 7.33–7.35 (m, 1H), 7.37–7.43 (m, 4H), 7.46 (d, J = 8.0 Hz, 2H); ^{13}C NMR (125 MHz, CDCl$_3$) δ: 55.3, 70.5, 72.8, 77.9, 79.3, 82.9, 106.0, 112.5, 113.7, 114.1 (2C), 116.9, 124.9, 127.6 (2C), 128.1, 128.6 (2C), 134.0 (2C), 136.7, 140.5, 145.6, 160.3; HRMS (FAB) calcd for C$_{24}$H$_{20}$NO$_2$ (MH$^+$): 354.1489, found 354.1494.

2-Azido-1-(benzyloxy)-3-[(4-methoxyphenyl)buta-1,3-diyn-1-yl]benzene (7d). To a solution of **22d** (2.05 g, 5.80 mmol) in MeCN (11.6 mL) were successively added *t*-BuONO (1.04 mL, 8.70 mmol) and TMSN$_3$ (0.930 mL, 7.01 mmol) at 0 °C. The mixture was stirred at room temperature for 2 h and concentrated in vacuo. The residue was purified by column chromatography (CHCl$_3$/hexane = 1/2) to give **7d** (1.72 g, 83%) as an off-white solid: mp 111 °C; IR (neat) 2134 (C≡C), 2093 (C≡C); ^{1}H NMR (500 MHz, CDCl$_3$) δ: ^{1}H-NMR (CDCl$_3$) δ: 3.83 (s, 3H), 5.15 (s, 2H), 6.84–6.87 (m, 2H), 6.94 (dd, J = 8.0, 1.4 Hz, 1H), 6.99 (dd, J = 8.0, 8.0 Hz, 1H), 7.08 (dd, J = 8.0, 1.4 Hz, 1H), 7.33–7.36 (m, 1H), 7.39–7.44 (m, 4H), 7.46–7.49 (m, 2H); ^{13}C NMR (125 MHz, CDCl$_3$) δ: 55.3, 71.4, 72.6, 77.2, 79.2,

83.5, 113.6, 113.9, 114.1 (2C), 115.4, 124.9, 126.2, 127.6 (2C), 128.3, 128.6 (2C), 131.2, 134.1 (2C), 135.5, 152.4, 160.4; HRMS (FAB) calcd for $C_{24}H_{17}N_3NaO_2$ (MNa$^+$): 402.1213, found 402.1216.

2.1.3 Gold-Catalyzed Cascade Cyclization

4-Phenyl-3,6-dihydropyrrolo[2,3-c]carbazole (9aa) and its [3,2-c]-isomer (10aa) (Table 2.1, entry 1). To a solution of **7a** (24 mg, 0.10 mmol) and **8a** (0.035 mL, 0.50 mmol) in DCE (0.50 mL) was added [BrettPhosAu(MeCN)SbF$_6$] (5.1 mg, 5.0 μmol) at 80 °C. The mixture was stirred at 80 °C in preheated bath for 8 h and concentrated in vacuo. The residue was purified by column chromatography (hexane/EtOAc = 5/1) to give a mixture of **9aa** and **10aa** containing a small amount of impurities (17.6 mg, <62%; **9aa:10aa** = 25:75). These isomers were separated by column chromatography on amine silica gel (hexane/CHCl$_3$ = 2/1) to give, in the order of elution, **9aa** and **10aa**.

Compound **9aa**: brown amorphous solid: IR (neat) 3406 (NH); ^{1}H NMR (500 MHz, CDCl$_3$) δ: 7.15 (dd, J = 2.5, 1.0 Hz, 1H), 7.24 (s, 1H), 7.30 (dt, J = 7.5, 1.0 Hz, 1H), 7.35 (t, J = 3.0 Hz, 1H), 7.38–7.45 (m, 3H), 7.52 (t, J = 8.0 Hz, 2H), 7.66–7.69 (m, 2H), 8.04 (s, 1H), 8.28 (d, J = 7.0 Hz, 1H), 8.54 (s, 1H); ^{13}C NMR (125 MHz, CDCl$_3$) δ: 101.4, 106.2, 110.5, 113.7, 119.2, 121.1, 121.3, 123.6, 124.18, 124.23, 125.0, 127.4, 128.5 (2C), 128.9, 129.2 (2C), 134.9, 139.1, 139.7; HRMS (ESI$^+$) calcd for $C_{20}H_{15}N_2$ (MH$^+$): 283.1230, found 283.1231.

Compound **10aa**: white solid: mp 231–234 °C; IR (neat) 3395 (NH); ^{1}H NMR (500 MHz, CDCl$_3$) δ: 6.87 (dd, J = 3.5,1.0 Hz, 1H), 7.31–7.34 (m, 3H), 7.39–7.44 (m, 2H), 7.51–7.54 (m, 3H), 7.78–7.79 (m, 2H), 8.08 (d, J = 6.5 Hz, 1H), 8.27 (s, 1H), 8.79 (s, 1H); ^{13}C NMR (125 MHz, CDCl$_3$) δ: 103.4, 104.6, 106.8, 110.7, 119.5, 120.2, 120.3, 121.7 (2C), 124.3, 127.0, 128.5 (2C), 129.1 (2C), 130.2, 133.7, 137.1, 138.7, 141.6; HRMS (ESI$^+$) calcd for $C_{20}H_{15}N_2$ (MH$^+$): 283.1230, found 283.1231.

3-Benzyl-4-phenyl-3,6-dihydropyrrolo[2,3-c]carbazole (9ab) and its [3,2-c]-isomer (10ab) (Table 2.1, entry 2). To a solution of **7a** (24 mg, 0.10 mmol) and **8b** (79 mg, 0.50 mmol) in DCE (0.50 mL) was added [BrettPhosAu(MeCN)SbF$_6$] (5.1 mg, 5.0 μmol) at 80 °C. The mixture was stirred at 80 °C in preheated bath for 10 h and concentrated in vacuo. The residue was purified by column chromatography (hexane/EtOAc = 10/1) to give a **9ab** (4.2 mg, 11%) and **10ab** (19 mg, 51%) in the order of elution, **9ab** and **10ab**; (**9ab:10ab** = 18:82).

Compound **9ab**: yellow solid: mp 163–168 °C; IR (neat) 3728 (NH); ^{1}H NMR (500 MHz, CDCl$_3$) δ: 5.00 (s, 2H), 6.47 (m, 2H), 7.05 (s, 1H), 7.06–7.13 (m, 3H), 7.20 (d, J = 3.0 Hz, 1H), 7.22–7.28 (m, 5H), 7.30–7.35 (m, 2H), 7.41 (dt, J = 8.0, 1.0 Hz, 1H), 7.48 (d, J = 8.0 Hz, 1H), 8.06 (s, 1H), 8.30 (d, J = 8.0 Hz, 1H); ^{13}C NMR (125 MHz, CDCl$_3$) δ: 52.2, 100.2, 108.7, 110.5, 113.8, 119.2, 121.2, 123.4, 123.7, 124.2, 125.8 (2C), 126.6, 126.9, 127.1, 127.6 (2C), 128.2 (2C), 128.7, 129.9 (2C), 130.8, 134.0, 138.8, 139.2, 140.6; HRMS (ESI$^+$) calcd for $C_{27}H_{21}N_2$ (MH$^+$): 373.1699, found 373.1700

Compound **10ab**: white solid: mp 173–178 °C; IR (neat) 3728 (NH); ^{1}H NMR (500 MHz, CDCl$_3$) δ: 6.02 (s, 2H), 6.84 (d, J = 3.0 Hz, 1H), 7.00–7.04 (m, 2H), 7.24–7.33 (m, 7H), 7.39–7.42 (m, 2H), 7.50 (t, J = 8.0 Hz, 2H), 7.75 (d, J = 7.5 Hz, 2H), 8.01 (d, J = 8.5 Hz, 1H), 8.23 (s, 1H); ^{13}C NMR (125 MHz, CDCl$_3$) δ: 53.2, 103.6, 104.8, 106.9, 110.6, 119.3, 121.0, 121.7, 122.6, 123.9, 126.5 (2C), 126.6, 127.0, 127.5, 128.4 (2C), 128.9 (2C), 129.2 (2C), 132.6, 134.1, 138.0, 138.7 (2C), 141.3; HRMS (ESI$^+$) calcd for C$_{27}$H$_{21}$N$_2$ (MH$^+$): 373.1699, found 373.1702.

4-Phenyl-3-tosyl-3,6-dihydropyrrolo[2,3-c]carbazole (9ac) and its [3,2-c]-isomer (10ac) (Table 2.1, entry 3). To a solution of **7a** (24 mg, 0.10 mmol) and **8c** (0.11 g, 0.50 mmol) in TCE (0.50 mL) was added [BrettPhosAu(MeCN)SbF$_6$] (10 mg, 10 μmol) at 140 °C. The mixture was stirred at 140 °C in preheated bath for 30 min and concentrated in vacuo. The residue was purified by column chromatography (hexane/EtOAc = 5/1) to give a mixture of **9ac** and **10ac** (15 mg, 34%; **9ac**:**10ac** = 58:42). These isomers were separated by column chromatography on silica gel (toluene) to give, in the order of elution, **9ac** and **10ac**.

Compound **9ac**: brown amorphous solid: IR (neat) 3736 (NH); ^{1}H NMR (500 MHz, CDCl$_3$) δ: 2.24 (s, 3H), 6.98 (d, J = 8.5 Hz, 2H), 7.18–7.21 (m, 3H), 7.27–7.31 (m, 2H), 7.37–7.47 (m, 7H), 7.80 (d, J = 4.0 Hz, 1H), 8.13–8.14 (m, 2H); ^{13}C NMR (125 MHz, CDCl$_3$) δ: 21.5, 110.0, 110.8, 111.6, 114.1, 119.8, 121.1, 122.8, 125.4, 126.5 (2C), 127.0, 127.4, 127.7 (2C), 128.9, 129.1 (2C), 129.3 (2C), 130.2, 132.0, 134.4, 136.6, 139.6, 141.8, 144.0; HRMS (ESI$^+$) calcd for C$_{27}$H$_{21}$N$_2$O$_2$S (MH$^+$): 437.1318, found 437.1318.

Compound **10ac**: white amorphous solid: IR (neat) 3390 (NH); ^{1}H NMR (500 MHz, CDCl$_3$) δ: 2.24 (s, 3H), 6.81 (d, J = 3.5 Hz, 1H), 6.96 (d, J = 8.5 Hz, 2H), 7.27–7.30 (m, 3H), 7.36–7.39 (m, 3H), 7.43–7.44 (m, 6H), 8.30 (s, 1H), 8.87 (d, J = 8.0 Hz, 1H); ^{13}C NMR (125 MHz, CDCl$_3$) δ: 21.5, 109.3, 110.0, 112.6, 114.2, 119.2, 121.7, 125.5, 125.7, 125.9, 126.8 (2C), 127.3, 128.3, 128.6 (2C), 128.9 (2C), 129.0 (2C), 132.4, 133.1, 133.6, 139.6, 139.7, 140.1, 144.3; HRMS (ESI$^+$) calcd for C$_{27}$H$_{21}$N$_2$O$_2$S (MH$^+$): 437.1318, found 437.1317.

tert-**Butyl 4-phenylpyrrolo[2,3-c]carbazole-3(6H)-carboxylate (9ad) and its [3,2-c]-isomer (10ad)** (Table 2.1, entry 4). To a solution of **7a** (24 mg, 0.10 mmol) and **8d** (84 mg, 0.50 mmol) in DCE (0.50 mL) was added [BrettPhosAu(MeCN)SbF$_6$] (5.1 mg, 5.0 μmol) at 80 °C. The mixture was stirred at 80 °C in preheated bath for 1.5 h and concentrated in vacuo. The residue was purified by column chromatography (hexane/EtOAc = 10/1) to give a mixture of **9ad** and **10ad** (23 mg, 60%; **9ad**:**10ad** = 92:8). These isomers were separated by reverse-phase chromatography on silica gel (MeCN/0.1% TFA aq.) to give, in the order of elution, **9ad** and **10ad**.

Compound **9ad**: pale yellow solid: mp 173–178 °C; IR (neat) 3397 (NH), 1742 (C=O); ^{1}H NMR (500 MHz, CDCl$_3$) δ: 1.30 (s, 9H), 7.21 (d, J = 3.5 Hz, 1H), 7.29 (s, 1H), 7.32 (t, J = 8.0 Hz, 2H), 7.40–7.46 (m, 4H), 7.57 (d, J = 8.0 Hz, 2H), 7.73 (d, J = 4.0 Hz, 1H), 8.16 (s, 1H), 8.24 (d, J = 8.0 Hz, 1H); ^{13}C NMR (125 MHz, CDCl$_3$) δ: 27.6 (3C), 83.5, 105.5, 110.1, 110.7, 113.8, 119.5, 121.2, 123.2, 124.9, 125.9, 126.6, 127.4 (2C), 127.5, 128.5 (2C), 128.8, 129.1, 136.4, 139.5, 142.7, 149.6; HRMS (ESI$^+$) calcd for C$_{25}$H$_{23}$N$_2$O$_2$ (MH$^+$): 383.1754, found 383.1755.

Compound **10ad**: white amorphous solid: IR (neat) 3592 (NH), 1747 (C=O); ^{1}H NMR (500 MHz, CDCl$_3$) δ: 1.72 (s, 9H), 6.84 (d, J = 3.5 Hz, 1H), 7.28 (ddd, J = 8.0, 6.5, 1.5 Hz, 1H), 7.38–7.44 (m, 4H), 7.50 (t, J = 7.0 Hz, 2H), 7.56 (d, J = 4.0 Hz, 1H), 7.65 (d, J = 7.0 Hz, 2H), 7.98 (d, J = 8.5 Hz, 1H), 8.27 (s, 1H); ^{13}C NMR (125 MHz, CDCl$_3$) δ: 28.2 (3C), 83.6, 107.6, 108.1, 110.0, 110.4, 119.0, 122.1, 123.2, 124.7, 125.4, 125.9, 127.1, 128.5 (2C), 129.2 (2C), 129.9, 133.3, 139.2, 139.5, 140.8, 150.0; HRMS (ESI$^+$) calcd for C$_{25}$H$_{23}$N$_2$O$_2$ (MH$^+$): 383.1754, found 383.1754.

***tert*-Butyl   7-(*tert*-butoxy)-4-(4-methoxyphenyl)pyrrolo[2,3-*c*]carbazole-3(6*H*)-carboxylate(9bd) and its [3,2-*c*]-isomer (10bd)** (Table 2.1, entry 10). To a solution of **7b** (35 mg, 0.10 mmol) and **8d** (84 mg, 0.50 mmol) in DCE (0.50 mL) was added [BrettPhosAu(MeCN)SbF$_6$] (5.1 mg, 5.0 μmol) at 80 °C. The mixture was stirred at 80 °C in preheated bath for 5 min and concentrated in vacuo. The residue was purified by column chromatography (hexane/EtOAc = 5/1) to give a mixture of **9bd** and **10bd** (39 mg, 79%; **9bd**:**10bd** = 84:16). These isomers were separated by column chromatography on silica gel (CHCl$_3$/Et$_2$O = 50/1) to give, in the order of elution, **10bd** and **9bd**. The reaction in gram scale using decreased amount of the catalyst also worked well to produce **9bd** (2.3 g, 58%), using **7** (2.8 g, 8.0 mmol), **8** (6.7 g, 40 mmol), and BrettPhosAu(MeCN)SbF$_6$ (0.16 g, 0.16 mmol) in DCE (40 mL).

Compound **9bd**: white amorphous solid: mp 114–115 °C; IR (neat) 3404 (NH), 1731 (C=O); ^{1}H NMR (500 MHz, CDCl$_3$) δ: 1.32 (s, 9H), 1.50 (s, 9H), 3.85 (s, 3H), 6.98 (d, J = 8.6 Hz, 2H), 7.10 (d, J = 7.4 Hz, 1H), 7.17–7.20 (m, 2H), 7.34 (s, 1H), 7.51 (d, J = 8.6 Hz, 2H), 7.71 (d, J = 3.4 Hz, 1H), 7.93 (d, J = 8.0 Hz, 1H), 8.29 (br s, 1H); ^{13}C NMR (125 MHz, CDCl$_3$) δ: 27.6 (3C), 29.2 (3C), 55.4, 80.0, 83.3, 105.4, 110.0, 113.97 (2C), 114.03, 116.0, 117.9, 119.5, 124.8, 125.9, 127.6, 128.4 (2C), 128.5, 129.1, 134.8, 135.4, 136.2, 140.8, 149.6, 158.5; HRMS (FAB) calcd for C$_{30}$H$_{33}$N$_2$O$_4$ (MH$^+$) 485.2435, found 485.2440.

Compound **10bd**: yellow oil: IR (neat) 3398 (NH), 1739 (C=O): ^{1}H NMR (500 MHz, CDCl$_3$) δ: 1.51 (s, 9H), 1.73 (s, 9H), 3.89 (s, 3H), 6.82 (d, J = 3.4 Hz, 1H), 7.05 (d, J = 8.3 Hz, 2H), 7.10 (d, J = 7.4 Hz, 1H), 7.14–7.18 (m, 1H), 7.43 (s, 1H), 7.54 (d, J = 3.4 Hz, 1H), 7.59 (d, J = 8.3 Hz, 2H), 7.69 (d, J = 8.0 Hz, 1H), 8.44 (br s, 1H); ^{13}C NMR (125 MHz, CDCl$_3$) δ: 28.2 (3C), 29.2 (3C), 55.3, 80.0, 83.5, 107.5, 108.1, 110.6, 114.0 (2C), 117.7, 118.8, 120.8, 123.1, 123.8, 125.1, 129.9, 130.2 (2C), 132.9, 133.3, 135.0, 139.1, 140.1, 150.0, 158.9; HRMS (FAB) calcd for C$_{30}$H$_{33}$N$_2$O$_4$ (MH$^+$) 485.2435, found 485.2443.

***tert*-Butyl   4-(4-methoxyphenyl)-7-[(methylsulfonyl)oxy]pyrrolo[2,3-*c*]carbazole-3(6*H*)-carboxylate (9cd) and its [3,2-*c*]-isomer (10cd)** (Table 2.1, entry 11). To a solution of **7c** (37 mg, 0.10 mmol) and **8d** (84 mg, 0.50 mmol) in DCE (0.50 mL) was added [BrettPhosAu(MeCN)SbF$_6$] (5.1 mg, 5.0 μmol) at 80 °C. The mixture was stirred at 80 °C in preheated bath for 5 min and concentrated in vacuo. The residue was purified by column chromatography (hexane/EtOAc = 2/1) to give a mixture of **9cd** and **10cd** (42 mg, 83%; **9cd**:**10cd** = 75:25). These isomers were separated by reverse-column chromatography (MeCN/0.1% TFA aq.) to give, in the order of elution, **9cd** and **10cd**.

Compound **9cd**: brown amorphous solid: mp 112–116 °C; IR (neat) 3422 (NH), 1720 (C=O); ^{1}H NMR (500 MHz, CDCl$_3$) δ: 1.33 (s, 9H), 3.26 (s, 3H), 3.58 (s, 3H), 6.98 (d, $J = 9.0$ Hz, 2H), 7.13 (d, $J = 3.5$ Hz, 1H), 7.24–7.32 (m, 3H), 7.48 (d, $J = 8.5$ H, 2H), 7.71 (d, $J = 3.0$ Hz, 1H), 8.12 (d, $J = 7.5$ Hz, 1H), 8.82 (s, 1H); ^{13}C NMR (125 MHz, CDCl$_3$) δ: 27.6 (3C), 37.1, 55.4, 83.6, 105.1, 110.2, 113.3, 114.0 (2C), 117.3, 119.6, 120.3, 125.7, 127.0, 128.0, 128.4 (2C), 129.4, 129.6, 132.2, 134.0, 135.0, 137.0, 149.4, 158.6; HRMS (ESI) calcd for C$_{27}$H$_{27}$N$_2$O$_6$S (MH$^+$): 507.1584, found 507.1581.

Compound **10cd**: off-white solid: mp 122–125 °C; IR (neat) 3398 (NH), 1736 (C=O); ^{1}H NMR (500 MHz, CDCl$_3$) δ: 1.73 (s, 9H), 3.23 (s, 3H), 3.89 (s, 3H), 6.83 (d, $J = 4.0$ Hz, 1H), 7.03–7.06 (m, 2H), 7.24 (t, $J = 8.0$ Hz, 1H), 7.32 (d, $J = 7.0$ Hz, 1H), 7.41 (s, 1H), 7.55–7.58 (m, 3H), 7.90 (d, $J = 8.0$ Hz, 1H), 8.94 (s, 1H); ^{13}C NMR (125 MHz, CDCl$_3$) δ: 28.2 (3C), 37.0, 55.4, 83.8, 107.8, 108.3, 109.9, 114.0 (2C), 117.0, 118.9, 123.8, 125.4, 125.5, 125.9, 129.7, 130.2 (2C), 132.3, 132.9, 133.7, 134.2, 139.8, 150.0, 159.1; HRMS (ESI) calcd for C$_{27}$H$_{27}$N$_2$O$_6$S (MH$^+$): 507.1584, found 507.1583.

***tert*-Butyl   7-(benzyloxy)-4-(4-methoxyphenyl)pyrrolo[2,3-*c*]carbazole-3(6*H*)-carboxylate (9dd) and its [3,2-*c*]-isomer (10dd)** (Table 2.1, entry 12). To a solution of **7d** (38 mg, 0.10 mmol) and **8d** (83 mg, 0.50 mmol) in DCE (0.16 mL) was added [BrettPhosAu(MeCN)SbF$_6$] (5.1 mg, 5.0 μmol) at 80 °C. The mixture was stirred at 80 °C in preheated bath for 5 min and concentrated in vacuo. The residue was purified by column chromatography (hexane/EtOAc = 5/1) to give a mixture of **9dd** and **10dd** (35 mg, 68%; **9dd**:**10dd** = 81:19). These isomers were separated by column chromatography on silica gel (CHCl$_3$/hexane = 6/1) to give, in the order of elution, **9dd** and **10dd**.

Compound **9dd**: white brown solid: mp 200 °C; IR (neat) 3368 (NH), 1749 (C=O); ^{1}H NMR (500 MHz, CDCl$_3$) δ: 1.33 (s, 9H), 3.85 (s, 3H), 5.28 (s, 2H), 6.98–7.00 (m, 3H), 7.18–7.22 (m, 2H), 7.30 (s, 1H), 7.38 (dd, $J = 7.2, 7.2$ Hz, 1H), 7.43 (dd, $J = 7.4, 7.4$ Hz, 2H), 7.50–7.54 (m, 4H), 7.70 (d, $J = 3.4$ Hz, 1H), 7.85 (d, $J = 8.0$ Hz, 1H), 8.40 (br s, 1H); ^{13}C NMR (125 MHz, CDCl$_3$) δ: 27.6 (3C), 55.4, 70.4, 83.3, 105.4, 106.3, 110.1, 113.92, 113.97 (2C), 114.03, 119.7, 124.3, 125.9, 127.7, 127.9 (2C), 128.2, 128.4 (2C), 128.5, 128.6 (2C), 129.0, 130.0, 135.4, 136.2, 136.9, 144.9, 149.6, 158.5; HRMS (FAB) calcd for C$_{33}$H$_{31}$N$_2$O$_4$ (MH$^+$): 519.2278, found 519.2281.

Compound **10dd**: white amorphous solid: IR (neat) 3401 (NH), 1721 (C=O); ^{1}H NMR (500 MHz, CDCl$_3$) δ: 1.72 (s, 9H), 3.88 (s, 3H), 5.78 (s, 2H), 6.82 (d, $J = 3.5$ Hz, 1H), 6.97 (d, $J = 8.0$ Hz, 1H), 7.04 (d, $J = 8.5$ Hz, 2H), 7.17 (t, $J = 8.0$ Hz, 1H), 7.35–7.39 (m, 2H), 7.43 (dd, $J = 7.0, 7.0$ Hz, 2H), 7.52–7.55 (m, 3H), 7.58 (d, $J = 8.0$ Hz, 2H), 7.62 (d, $J = 8.0$ Hz, 1H), 8.56 (s, 1H); ^{13}C NMR (125 MHz, CDCl$_3$) δ: 28.1 (3C), 55.3, 70.5, 83.5, 106.1, 107.6, 108.1, 110.5, 114.0 (2C), 118.7, 118.9, 123.1, 123.3, 125.2, 127.8 (2C), 128.1, 128.6 (2C), 130.0, 130.1, 130.2 (2C), 132.9, 133.3, 137.1, 139.0, 144.3, 150.0, 158.9; HRMS (FAB) calcd for C$_{33}$H$_{31}$N$_2$O$_4$ (MH$^+$): 519.2278, found 519.2281.

2.1.4 Total and Formal Synthesis of Dictyodendrins A–F

7-(*tert*-Butoxy)-4-(4-methoxyphenyl)-3,6-dihydropyrrolo[2,3-*c*]carbazole (24). To a solution of **9bd** (2.07 g, 4.27 mmol) in THF (45 mL) was added 5 M NaOMe (4.30 mL, 21.4 mmol) at room temperature under Ar. The mixture was stirred at 50 °C for 4.5 h. The mixture was diluted with saturated aqueous NH_4Cl, and the aqueous layer was extracted twice with Et_2O. The combined organic layer was washed with H_2O and brine, dried over Na_2SO_4, filtered, and concentrated in vacuo. The residue was purified by column chromatography (hexane/EtOAc = 4/1) to give **24** (1.51 g, 92%) as an white amorphous solid: IR (neat) 3353 (NH); ^{1}H NMR (500 MHz, CDCl$_3$) δ: 1.51 (s, 9H), 3.88 (s, 3H), 7.05–7.07 (m, $J = 9.0$ Hz, 2H), 7.09 (d, $J = 8.0$ Hz, 1H), 7.13 (dd, $J = 2.0, 2.0$ Hz, 1H), 7.18 (dd, $J = 8.5, 8.5$ Hz, 1H), 7.30 (s, 1H), 7.35 (dd, $J = 3.0, 3.0$ Hz, 1H), 7.61–7.63 (d, $J = 8.5$ Hz, 2H), 7.98 (d, $J = 7.5$ Hz, 1H), 8.26 (s, 1H), 8.53 (s, 1H); ^{13}C NMR (125 MHz, CDCl$_3$) δ: 29.2 (3C), 55.4, 79.9, 101.3, 106.1, 113.8, 114.6 (2C), 116.0, 117.3, 119.1, 121.3, 124.0, 124.7, 125.2, 129.0, 129.5 (2C), 132.1, 134.4, 134.6, 140.7, 159.0; HRMS (FAB) calcd for $C_{25}H_{25}N_2O_2$ (MH$^+$): 385.1911, found 385.1919.

1-Bromo-7-(*tert*-butoxy)-4-(4-methoxyphenyl)-3,6-dihydropyrrolo[2,3-*c*]carbazole (25). To a solution of **24** (510 mg, 1.33 mmol) and K_2CO_3 (920 mg, 6.65 mmol) in dioxane (30 mL) was added NBS (260 mg, 1.46 mmol) at room temperature. The mixture was stirred at 75 °C under Ar for 17 h. The mixture was cooled to room temperature and diluted with saturated brine, and the aqueous layer was extracted twice with EtOAc. The combined organic layer was dried over Na_2SO_4, filtered, and concentrated in vacuo. The residue was purified by column chromatography (hexane/EtOAc = 4/1) to give **25** (446 mg, 72%) as an amorphous solid: IR (neat) 3356 (NH); ^{1}H NMR (500 MHz, CDCl$_3$) δ: 1.52 (s, 9H), 3.99 (s, 3H), 7.06 (m, 3H), 7.15 (t, $J = 8.0$ Hz, 1H), 7.33 (m, 2H), 7.55–7.57 (m, 2H), 8.39 (s, 1H), 8.55 (s, 1H), 9.03 (d, $J = 8.0$ Hz, 1H); ^{13}C NMR (125 MHz, CDCl$_3$) δ: 29.2 (3C), 55.4, 79.9, 89.6, 107.5, 113.2, 114.7(2C), 117.0, 118.9, 120.3, 121.3, 124.1, 124.4, 125.0, 129.4, 129.6 (2C), 130.9, 134.4, 135.3, 140.5, 159.3; HRMS (ESI$^+$) calcd for $C_{25}H_{24}BrN_2O_2$ (MH$^+$): 463.1016, found 463.1017.

7-(*tert*-Butoxy)-1,4-bis(4-methoxyphenyl)-3,6-dihydropyrrolo[2,3-*c*]carbazole (27). To a solution of **25** (100 mg, 0.216 mmol), **26** (320 mg, 2.16 mmol), and K_3PO_4 (920 mg, 4.32 mmol) in dioxane/H_2O (10/1, 5.0 mL) was added Pd(*t*-Bu$_3$P)$_2$ (11.0 mg, 0.0216 mmol) at room temperature. The mixture was stirred at 80 °C under Ar for 6 h. The mixture was cooled to room temperature and diluted with brine, and the aqueous layer was extracted twice with EtOAc. The combined organic layer was washed with H_2O and brine, dried over Na_2SO_4, filtered, and concentrated in vacuo. The residue was purified by column chromatography (hexane/EtOAc = 3/1) to give **27** (108 mg, quant) as an off-white solid: mp 190–193 °C; IR (neat) 3447 (NH); ^{1}H NMR (600 MHz, CDCl$_3$) δ: 1.49 (s, 9H), 3.92 (s, 3H), 3.94 (s, 3H), 6.46 (d, $J = 7.0$ Hz, 1H), 6.75 (t, $J = 7.0$ Hz, 1H), 6.94 (d, $J = 5.5$ Hz, 1H), 7.04–7.06 (m, 2H), 7.10–7.11 (m, 2H), 7.26 (m, 1H), 7.36 (s, 1H), 7.53–7.54 (m, 2H), 7.66–7.68 (m, 2H), 8.33 (s, 1H), 8.56(s, 1H); ^{13}C NMR

(150 MHz, CDCl$_3$) δ: 29.2 (3C), 55.4, 55.5, 79.8, 106.6, 113.4 (2C), 114.0, 114.6 (2C), 116.7, 118.3, 119.0, 119.3, 121.1, 122.9, 124.7, 124.8, 129.3, 129.7 (2C), 130.1, 131.8, 132.1 (2C), 134.5, 135.3, 140.2, 158.9, 159.1; HRMS (ESI$^+$) calcd for C$_{32}$H$_{31}$N$_2$O$_3$ (MH$^+$): 491.2324, found 491.2326.

1-Bromo-7-(*tert*-butoxy)-3-(4-methoxyphenethyl)-4-(4-methoxyphenyl)-3,6-dihydropyrrolo[2,3-*c*]carbazole (29) (Table 2.2, entry 7). To a solution of **25** (100 mg, 0.216 mmol) in THF/H$_2$O (10/1, 4.5 mL) were added (C$_2$H$_4$O)$_6$ (171 mg, 0.648 mmol), **28** (0.340 mL, 2.18 mmol), and NaOH (130 mg, 3.24 mmol) at 0 °C. The mixture was warmed to room temperature and stirred at room temperature for 18 h and diluted with saturated aqueous NH$_4$Cl, and the aqueous layer was extracted twice with EtOAc. The combined organic layer was washed with H$_2$O and brine, dried over Na$_2$SO$_4$, filtered, and concentrated in vacuo. The residue was filtered through by short column chromatography (hexane/EtOAc = 3/1) to give **29** (106 mg, 82%) as an off-white solid: mp 150–153 °C; IR (neat) 3364 (NH); ^{1}H NMR (600 MHz, CDCl$_3$) δ: 1.51 (s, 9H), 2.51–2.57 (m, 2H), 3.75 (s, 3H), 3.89–3.95 (m, 5H), 6.54–6.56 (m, 2H), 6.68–6.71 (m, 2H), 7.02–7.04 (m, 2H), 7.07–7.10 (m, 2H), 7.15 (t, J = 7.5 Hz, 1H), 7.19 (s, 1H), 7.45–7.46 (m, 2H), 8.33 (s, 1H), 9.09 (d, J = 8.0 Hz, 1H); ^{13}C NMR (150 MHz, CDCl$_3$) δ: 29.2 (3C), 36.7, 50.8, 55.3, 55.5, 79.9, 87.7, 110.5, 113.6 (3C), 113.7 (2C), 117.2, 118.9, 120.7, 123.4, 124.4, 126.1, 129.0, 129.6 (2C), 129.9, 130.4, 130.9 (2C), 132.9, 134.6, 134.7, 140.5, 158.2, 159.3; HRMS (ESI$^+$) calcd for C$_{34}$H$_{34}$BrN$_2$O$_3$ (MH$^+$): 597.1747, found 597.1743.

7-(*tert*-Butoxy)-3-(4-methoxyphenethyl)-1,4-bis(4-methoxyphenyl)-3,6-dihydropyrrolo[2,3-*c*]carbazole (14a). To a solution of **24** (320 mg, 0.832 mmol) in THF (20 mL) was added NBS (156 mg, 0.874 mmol) at 0 °C. The mixture was stirred at 0 °C for 1 h. To the mixture were added H$_2$O (1.5 mL), (C$_2$H$_4$O)$_6$ (770 mg, 29.1 mmol), **28a** [32] (1.30 mL, 8.30 mmol), and NaOH (500 mg, 12.5 mmol) at 0 °C. The mixture was warmed to room temperature and stirred at room temperature. After 54 h, the second portion of **28a** (0.650 mL, 4.15 mmol) and NaOH (250 mg, 6.25 mmol), and after 24 h, the third portion of **28a** (0.260 mL, 1.67 mmol) and NaOH (100 mg, 1.67 mmol) were added at room temperature. The mixture was stirred at room temperature for additional 6 h and diluted with saturated aqueous NH$_4$Cl, and the aqueous layer was extracted twice with EtOAc. The combined organic layer was washed with H$_2$O and brine, dried over Na$_2$SO$_4$, filtered, and concentrated in vacuo. The residue was filtered through by short column chromatography (hexane/EtOAc = 3/1) to give crude **29** (254 mg) as an off-white solid. To a solution of crude **29** (254 mg), **26** (650 mg, 4.25 mmol), and K$_3$PO$_4$ (1.80 g, 8.48 mmol) in dioxane/H$_2$O (10/1, 11.0 mL) was added Pd(*t*-Bu$_3$P)$_2$ (22.0 mg, 0.0430 mmol) at room temperature. The mixture was stirred at 80 °C under Ar for 14 h. The mixture was cooled to room temperature and diluted with brine, and the aqueous layer was extracted twice with EtOAc. The combined organic layer was washed with H$_2$O and brine, dried over Na$_2$SO$_4$, filtered, and concentrated in vacuo. The residue was purified by column chromatography (hexane/EtOAc = 3/1) to give **14a** (212 mg, 42% in 3 steps) as an off-white solid: mp 215–218 °C; IR (neat) 3537 (NH); ^{1}H NMR (500 MHz, CDCl$_3$) δ: 1.47 (s, 9H), 2.55–2.58 (m, 2H), 3.73 (s, 3H), 3.89 (s, 3H), 3.92 (s, 3H), 3.94–3.97 (m, 2H), 6.37 (d, J = 8.0 Hz, 1H),

6.54–6.56 (m, 2H), 6.67–6.69 (m, 2H), 6.72 (dd, $J = 8.0$, 8.0 Hz, 1H), 6.93–6.94 (m, 2H), 7.01–7.03 (m, 4H), 7.19 (s, 1H), 7.45–7.47 (m, 2H), 7.50–7.53 (m, 2H), 8.27 (s, 1H); ^{13}C NMR (125 MHz, CDCl$_3$) δ: 29.1 (3C), 36.7, 50.6, 55.2, 55.41, 55.42, 79.7, 109.4, 113.3 (2C), 113.5 (2C), 113.6 (2C), 114.3, 116.8, 117.1, 118.2, 119.6, 123.2, 124.7, 125.9, 128.6, 129.55, 129.59 (2C), 130.1, 130.4, 131.0 (2C), 132.1 (2C), 133.4, 134.5, 134.6, 140.1, 158.0, 158.8, 159.1; HRMS (FAB) calcd for C$_{41}$H$_{41}$N$_2$O$_2$ (MH$^+$): 625.3061, found 625.3066.

2,5-Dibromo-7-(*tert*-butoxy)-3-(4-methoxyphenethyl)-1,4-bis(4-methoxyphenyl)-3,6-dihydropyrrolo[2,3-*c*]-carbazole (30). To a solution of **14a** (103 mg, 0.165 mmol) in THF (3.3 mL) was added NBS (60.2 mg, 0.388 mmol) at −78 °C and the mixture was stirred at −78 °C under Ar for 45 min. The mixture was allowed to warm to room temperature and stirred for 3 h. The mixture was diluted with saturated aqueous NH$_4$Cl, and the aqueous layer was extracted twice with Et$_2$O. The combined organic layer was washed with H$_2$O and brine, dried over Na$_2$SO$_4$, filtered, and concentrated in vacuo. The residue was purified by column chromatography (hexane/CHCl$_3$ = 1/1) to give **30** (70.0 mg, 54%) as an off-white solid: mp 216–218 °C; IR (neat) 3475 (NH); ^{1}H NMR (500 MHz, CDCl$_3$) δ: 1.47 (s, 9H), 2.57–2.61 (m, 2H), 3.76 (s, 3H), 3.90–3.94 (m, 8H), 5.97 (d, $J = 8.0$ Hz, 1H), 6.68–6.73 (m, 5H), 6.96 (d, $J = 8.0$ Hz, 1H), 7.07–7.10 (m, 4H), 7.41–7.45 (m, 4H), 8.47 (s, 1H); ^{13}C NMR (125 MHz, CDCl$_3$) δ: 29.1 (3C), 35.7, 48.3, 55.2, 55.35, 55.43, 80.1, 104.5, 113.6 (2C), 113.8 (2C), 113.89, 113.93 (2C), 116.0, 117.6, 117.7, 119.0, 119.2, 122.4, 125.0, 128.3, 128.6, 129.0, 129.6 (2C), 129.91, 129.94, 131.8 (2C), 133.0 (2C), 133.3, 134.1, 140.4, 158.2, 159.4, 159.6; HRMS (FAB) calcd for C$_{41}$H$_{39}$Br$_2$N$_2$O$_4$ (MH$^+$): 781.1271, found 781.1270.

5-Bromo-7-(*tert*-butoxy)-3-(4-methoxyphenethyl)-1,4-bis(4-methoxyphenyl)-3,6-dihydropyrrolo[2,3-*c*]-carbazole (31) [33]. To a solution of **30** (41 mg, 0.049 mmol) and PdCl$_2$(dppf) (20 mg, 0.024 mmol) in THF (2.0 mL) were added TMEDA (73μL, 0.49 mmol) and NaBH$_4$ (18 mg, 0.49 mmol) at room temperature. The mixture was stirred at 55 °C under Ar for 0.5 h. The mixture was diluted with saturated aqueous NH$_4$Cl, and the aqueous layer was extracted twice with EtOAc. The combined organic layer was washed with H$_2$O and brine, dried over Na$_2$SO$_4$, filtered, and concentrated in vacuo. The residue was purified by column chromatography (hexane/EtOAc = 4/1) to give **31** (23 mg, 55%) as an off-white solid: mp 202–205 °C; IR (neat) 3448 (NH); ^{1}H NMR (500 MHz, CDCl$_3$) δ: 1.49 (s, 9H), 2.62–2.65 (m, 2H), 3.75–3.79 (m, 5H), 3.93 (s, 6H), 6.33 (d, $J = 8.0$ Hz, 1H), 6.65 (d, $J = 8.5$ Hz, 2H), 6.71–6.76 (m, 3H), 6.88 (s, 1H), 6.97 (d, $J = 8.0$ Hz, 1H), 7.02 (d, $J = 8.5$ Hz, 2H), 7.09 (d, $J = 8.0$ Hz, 2H), 7.42–7.46 (m, 4H), 8.48 (s, 1H); ^{13}C NMR (125 MHz, CDCl$_3$) δ: 29.2 (3C), 37.1, 50.0, 55.2, 55.3, 55.5, 80.0, 103.9, 113.4 (2C), 113.6 (2C), 113.7 (2C), 114.7, 116.8, 117.5, 118.7, 119.8, 122.1, 124.8, 125.3, 129.57, 129.64 (2C), 129.7, 129.8, 130.3, 131.3, 131.9 (2C), 132.1 (2C), 133.1, 134.1, 140.4, 158.2, 158.9, 159.5; HRMS (ESI$^+$) calcd for C$_{41}$H$_{40}$BrN$_2$O$_4$ (MH$^+$): 703.2166, found 703.2169.

7-(*tert*-Butoxy)-5-methoxy-3-(4-methoxyphenethyl)-1,4-bis(4-methoxyphenyl)-3,6-dihydropyrrolo[2,3-*c*]carbazole (13a): formal synthesis of dictyodendrin C [40]. To a solution of **31** (7.6 mg, 0.011 mmol) and CuI

(6.2 mg, 0.032 mmol) in DMF (0.20 mL) was added 4 M NaOMe in MeOH, separately prepared in a loosely capped vial by the portionwise addition of sodium metal (92 mg, 4.0 mmol) to vigorously stirred MeOH (1.0 mL). The mixture was stirred at 80 °C in preheated bath under Ar for 11 h. The mixture was diluted with saturated aqueous NH_4Cl, and the aqueous layer was extracted twice with EtOAc. The combined organic layer was washed with H_2O and brine, dried over Na_2SO_4, filtered, and concentrated in vacuo. The residue was purified by column chromatography (hexane/EtOAc = 4/1) to give **13a** (7.3 mg, quant) as an off-white solid: mp 220–223 °C; IR (neat) 3431 (NH); ^1H NMR (500 MHz, $CDCl_3$) δ: 1.48 (s, 9H), 2.59–2.63 (m, 2H), 3.64 (s, 3H), 3.75 (s, 3H), 3.83–3.86 (m, 2H), 3.931 (s, 3H), 3.936 (s, 3H), 6.38 (d, J = 8.0 Hz, 1H), 6.62 (m, 2H), 6.70–6.75 (m, 3H), 6.90 (s, 1H), 6.95 (d, J = 8.5 Hz, 1H), 7.01–7.03 (m, 2H), 7.08–7.09 (m, 2H), 7.45–7.47 (m, 2H), 7.54–7.56 (m, 2H), 8.35 (s, 1H); ^{13}C NMR (125 MHz, $CDCl_3$) δ: 29.2 (3C), 37.0, 50.2, 55.2, 55.4, 55.5, 61.3, 79.9, 113.3, 113.6, 113.7, 115.3, 116.9, 117.2, 117.8, 118.4, 119.5, 119.6, 125.4, 127.6, 128.7, 128.8, 129.0, 129.6 (2C), 130.1, 130.5, 132.1 (2C), 132.3 (2C), 134.4, 140.2, 140.4, 158.1, 158.8, 159.3; HRMS (ESI⁺) calcd for $C_{42}H_{43}N_2O_5$ (MH⁺): 655.3166, found 655.3164.

Dictyodendrin F (6) [41]. To a solution of **13a** (9.6 mg, 0.015 mmol) and cyclohexene (30 μL, 0.29 mmol) in CH_2Cl_2 (0.20 mL) was added BBr_3 (1 M in CH_2Cl_2, 0.15 mL, 0.15 mmol) dropwise at −78 °C and resulting mixture was allowed to reach ambient temperature over 1.5 h under Ar. The mixture was stirred at room temperature for 1.5 h and diluted with aq. $KHSO_4$ (10% w/w, 2 mL) and NaOH (20% w/w, 1 mL). The organic phase was washed with water. The aqueous layer was acidified with conc. HCl (2.0 mL) and extracted twice with *tert*-butyl methyl ether. The combined organic layer was washed with brine, dried over Na_2SO_4, filtered, and concentrated in vacuo. The residue was purified by reverse-phase column chromatography (MeCN/0.1% TFA aq.) to give dictyodendrin F (**6**) (3.4 mg, 42%) as a green brown amorphous: IR (neat) 3595 (OH), 1684 (C=O); ^1H NMR (500 MHz, CD_3OD) δ: 2.39–2.42 (m, 2H), 3.41–3.44 (m, 2H), 5.83 (dd, J = 7.5, 1.5 Hz, 1H), 6.56–6.59 (m, 4H), 6.66 (d, J = 8.0 Hz, 2H), 6.90 (d, J = 8.5 Hz, 2H), 6.94 (d, J = 8.5 Hz, 2H), 7.25 (d, J = 9.0 Hz, 2H), 7.31 (d, J = 8.5 Hz, 2H); ^{13}C NMR (125 MHz, CD_3OD) δ: 34.9, 44.1, 110.0, 114.0, 116.0 (2C), 116.1 (2C), 116.2 (2C), 116.4, 119.3, 123.1, 123.9, 124.2, 126.3, 130.0, 130.1, 130.7, 130.9 (2C), 132.2, 133.5 (2C), 133.9 (2C), 135.6, 146.0, 150.5, 156.9, 159.2, 160.1, 173.4, 180.9; HRMS (ESI⁺) calcd for $C_{34}H_{25}N_2O_6$ (MH⁺): 557.1707, found 557.1712.

3-[4-(Benzyloxy)phenethyl]-7-(*tert*-butoxy)-1,4-bis(4-methoxyphenyl)-3,6-dihydropyrrolo[2,3-*c*]carbazole (14b). To a solution of **24** (120 mg, 0.312 mmol) in THF (6.0 mL) was added NBS (58.2 mg, 0.327 mmol) at 0 °C. The mixture was stirred at 0 °C for 1 h. To the mixture were added H_2O (0.600 mL), $(C_2H_4O)_6$ (290 mg, 4.68 mmol), **28d** (906 mg, 3.12 mmol), and NaOH (190 mg, 4.68 mmol) at 0 °C. The mixture was warmed to room temperature and stirred at room temperature. After 48 h, the second portion of **28d** (227 mg, 0.780 mmol) and NaOH (31.2 mg, 0.780 mmol) was added at room temperature. The mixture was stirred at room temperature for additional 24 h and diluted with saturated aqueous NH_4Cl, and the aqueous layer was extracted twice with EtOAc. The combined organic layer was washed with H_2O and brine, dried over Na_2SO_4, filtered, and concentrated in vacuo. The residue was through by short column chromatography (hexane/EtOAc = 3/1) to give resulting crude **S5** as an off-white solid. To a solution of **S5**, **26** (474 mg, 3.12 mmol), and K_3PO_4 (1.30 g, 6.24 mmol) in dioxane/H_2O (10/1, 6.0 mL) was added Pd(t-Bu$_3$P)$_2$ (16.0 mg, 0.0312 mmol) at room temperature. The mixture was stirred at 80 °C under Ar for 2 h. The mixture was cooled to room temperature and diluted with brine, and the aqueous layer was extracted twice with EtOAc. The combined organic layer was washed with H_2O and brine, dried over Na_2SO_4, filtered, and concentrated in vacuo. The residue was purified by column chromatography (hexane/EtOAc = 3/1) to give **28** (107 mg, 49% in 3 steps) as an off-white solid: mp 197–199 °C; IR (neat) 3378 (NH); ^{1}H NMR (500 MHz, CDCl$_3$) δ: 1.50 (s, 9H), 2.58–2.61 (m, 2H), 3.92 (s 3H), 3.94 (s, 3H), 3.98–4.01 (m, 2H), 5.00 (s, 2H), 6.40 (d, J = 8.0 Hz, 1H), 6.58 (d, J = 8.0 Hz,2H), 6.73–6.79 (m, 3H), 6.96–6.97 (m, 2H), 7.04–7.06 (m, 4H), 7.21 (s, 1H), 7.32–7.35 (m, 1H), 7.38–7.43 (m, 4H), 7.48–7.51 (m,2H), 7.54–7.55 (m, 2H), 8.30 (s, 1H); ^{13}C NMR (125 MHz, CDCl$_3$) δ: 29.1(3C), 36.7, 50.5, 55.5(2C), 69.9, 79.7, 109.4, 113.3(2C), 113.5(2C), 114.4, 114.6, 116.8, 117.2, 118.2, 119.6, 123.2, 124.8, 125.9, 127.4(2C), 127.9, 128.5(2C), 128.6(2C), 129.5, 129.6(2C), 130.2, 130.7, 131.0(2C), 132.1(2C), 133.5, 134.5, 134.6, 137.0, 140.1, 157.3, 158.8, 159.1; HRMS (ESI$^+$) calcd for $C_{47}H_{45}N_2O_4$ (MH$^+$): 701.3374, found 701.3376.

NBS (2.05 eq.)

(53%)

14b   **S6**

3-[4-(Benzyloxy)phenethyl]-2,5-dibromo-7-(*tert*-butoxy)-1,4-bis(4-methoxyphenyl)-3,6-dihydropyrrolo[2,3-*c*]carbazole (S6). To a solution of **14b** (26.6 mg, 0.0380 mmol) in THF (1.0 mL) was added NBS (13.8 mg, 0.778 mmol) at 0 °C, and the mixture was stirred at 0 °C under Ar for 1 h. The mixture was diluted with saturated aqueous NH_4Cl, and the aqueous layer was extracted twice with Et_2O. The combined organic layer was washed with H_2O and brine, dried over Na_2SO_4, filtered, and concentrated in vacuo. The residue was purified by column chromatography (hexane/$CHCl_3$ = 1/1) to give **S6** (17.3 mg, 53%) as an off-white solid: mp 100–105 °C; IR (neat) 3471 (NH); ^1H NMR (500 MHz, $CDCl_3$) δ:1.49 (s, 9H), 2.59–2.63 (m, 2H), 3.91–3.96 (m, 8H), 5.02 (s, 2H), 5.97 (d, J = 8.0 Hz, 1H), 6.71 (t, J = 8.0 Hz,1H), 6.72–6.76 (m, 2H), 6.81–6.82 (m, 2H), 6.96 (d, J = 7.0 Hz, 1H), 7.09–7.12 (m, 3H), 7.32–7.35 (m, 1H), 7.38–7.46 (m, 8H), 8.47 (s, 1H); ^{13}C NMR (125 MHz, $CDCl_3$) δ: 29.1(3C), 35.8, 48.3, 55.36, 55.44, 70.0, 80.1, 104.5, 113.8 (2C), 113.9, 114.0 (2C), 114.6 (2C), 117.6, 117.7, 119.0, 119.2, 122.4, 124.5, 124.9, 127.4 (2C), 127.9 (2C), 128.6 (2C), 129.0, 129.7 (2C), 129.9, 130.2, 131.1, 131.8 (2C), 133.0 (2C), 133.3, 134.1, 137.0, 140.4, 157.4, 159.4, 159.7; HRMS (ESI$^+$) calcd for $C_{47}H_{43}Br_2N_2O_4$ (MH$^+$): 857.1584, found 857.1581.

3-[4-(Benzyloxy)phenethyl]-5-bromo-7-(*tert*-butoxy)-1,4-bis(4-methoxyphenyl)-3,6-dihydropyrrolo[2,3-*c*]carbazole (S7). To a solution of **S6** (21 mg, 0.024 mmol) and PdCl$_2$(dppf) (10 mg, 0.012 mmol) in THF (0.30 mL) were added TMEDA (28 mg, 0.24 mmol) and NaBH$_4$ (9.1 mg, 0.24 mmol) at room temperature. The mixture was stirred at 55 °C under Ar for 2 h. The mixture was diluted with saturated aqueous H_2O, and the aqueous layer was extracted twice with EtOAc. The combined organic layer was washed with brine, dried over Na_2SO_4, filtered, and concentrated in vacuo. The residue was purified by column chromatography (hexane/EtOAc = 4/1) to give **S7** (11 mg, 60%) as an off-white solid: mp 132–137 °C; IR (neat) 3475 (NH); ^1H NMR (500 MHz, $CDCl_3$) δ: 1.49 (s, 9H), 2.62–2.65 (m, 2H), 3.75–3.78 (m, 2H), 3.91 (s 3H), 3.91 (s, 3H), 5.00 (s, 2H), 6.32 (d, J = 8.0 Hz, 1H), 6.65 (d, J = 8.5 Hz, 2H), 6.75 (t, J = 8.0 Hz, 1H), 6.79 (d, J = 8.0 Hz, 2H), 6.88 (s, 1H), 6.96 (d, J = 8.0 Hz, 1H), 7.02 (d, J = 8.0 Hz, 1H), 7.08 (d, J = 8.0 Hz, 1H), 7.31–7.34 (m, 1H), 7.36–7.43 (m, 8H), 8.45 (s, 1H);

^{13}C NMR (125 MHz, CDCl$_3$) δ: 29.2 (3C), 37.1, 50.0, 55.3, 55.5, 70.0, 80.0, 104.0, 113.3 (2C), 113.8 (2C), 114.6 (2C), 117.5, 118.7, 119.9, 122.1, 124.8, 125.3, 127.4 (2C), 128.0, 128.6 (2C), 129.69 (2C), 129.73, 129.8, 130.2, 130.6, 131.3, 131.9 (2C), 132.1 (2C), 133.1, 134.1, 135.3, 137.0, 140.2, 140.4, 157.4, 158.9, 159.5; HRMS (ESI$^+$) calcd for C$_{47}$H$_{44}$BrN$_2$O$_4$ (MH$^+$): 779.2479, found 779.2470.

3-[4-(Benzyloxy)phenethyl]-7-(*tert*-butoxy)-5-methoxy-1,4-bis(4-methoxyphenyl)-3,6-dihydropyrrolo[2,3-*c*]carbazole (13b). To a solution of **S7** (12.7 mg, 0.0163 mmol) and CuI (9.3 mg, 0.0489 mmol) in DMF (300 µL) was added 4 M NaOMe in MeOH (120 µL), separately prepared in a loosely capped vial by the portionwise addition of sodium metal (92.0 mg, 4.00 mmol) to vigorously stirred MeOH (1.0 mL). The mixture was stirred at 80 °C in preheated bath under Ar for 4 h. The mixture was diluted with saturated aqueous NH$_4$Cl, and the aqueous layer was extracted twice with EtOAc. The combined organic layer was washed with H$_2$O and brine, dried over Na$_2$SO$_4$, filtered, and concentrated in vacuo. The residue was purified by column chromatography (hexane/EtOAc = 4/1) to give **13b** (5.04 mg, 43%) as an off-white solid: mp 188–190 °C; IR (neat) 3483 (NH); ^{1}H NMR (600 MHz, CDCl$_3$) δ: 1.48 (s, 9H), 2.60–2.62 (m, 2H), 3.64 (s, 3H), 3.85–3.86 (m, 2H), 3.92 (s, 3H), 3.93 (s, 3H), 5.00 (s, 2H), 6.37 (d, J = 7.8 Hz, 1H), 6.62–6.63 (m, 2H), 6.73 (t, J = 7.2 Hz, 1H), 6.77–6.79 (m, 2H), 6.89 (s, 1H), 6.95 (d, J = 6.6 Hz, 1H), 7.01–7.03 (m, 2H), 7.07–7.09 (m, 2H), 7.31–7.33 (m, 1H), 7.37–7.42 (m, 4H), 7.46–7.47 (m, 2H), 7.54–7.56 (m, 2H), 8.35 (s, 1H); ^{13}C NMR (150 MHz, CDCl$_3$) δ: 29.2(3C), 37.0, 50.2, 55.4, 55.5, 61.3, 70.0, 79.9, 113.3 (2C), 113.7 (2C), 114.6 (2C), 115.3, 117.2, 117.8, 118.4, 119.5, 119.6, 125.3, 127.5, 127.6, 128.0, 128.2, 128.6 (2C), 128.7, 128.8, 129.0, 129.1, 129.7 (2C), 130.1, 130.8, 132.1 (2C), 132.3 (2C), 134.4, 137.0, 140.2, 140.3, 157.3, 158.8, 159.3; HRMS (ESI$^+$) calcd for C$_{48}$H$_{47}$N$_2$O$_5$ (MH$^+$): 731.3479, found 731.3485.

2-Bromo-7-(*tert*-butoxy)-3-(4-methoxyphenethyl)-1,4-bis(4-methoxyphenyl)-3,6-dihydropyrrolo[2,3-*c*]carbazole (32). To a solution of **14a** (500 mg, 0.801 mmol) in THF (20 mL) was added NBS (153 mg, 0.861 mmol) at −78 °C. The mixture was stirred at −78 °C under Ar for 0.5 h. The mixture was diluted with saturated aqueous NH$_4$Cl, and the aqueous layer was extracted twice with EtOAc. The combined organic layer was washed with H_2O and brine, dried over Na_2SO_4, filtered, and concentrated in vacuo. The residue was purified by column chromatography (hexane/EtOAc = 3/1) to give **32** (291 mg, 52%) as an off-white solid: mp 200 °C (dec.); IR (neat) 3421 (NH); ^1H NMR (500 MHz, $CDCl_3$) δ: 1.46 (s, 9H), 2.51–2.55 (m, 2H), 3.75 (s, 3H), 3.92 (s, 3H), 3.96 (s, 3H), 4.11–4.15 (m, 2H), 6.00 (d, $J = 6.4$ Hz, 1H), 6.62–6.71 (m, 5H), 6.92 (d, $J = 7.0$ Hz, 1H), 7.05 (d, $J = 6.7$ Hz, 2H), 7.09 (d, $J = 6.7$ Hz, 2H), 7.18 (s, 1H), 7.46 (d, $J = 7.0$ Hz, 2H), 7.54 (d, $J = 7.0$ Hz, 2H), 8.25 (s, 1H); ^{13}C NMR (125 MHz, $CDCl_3$) δ: 29.2 (3C), 35.6, 48.6, 55.2, 55.4, 55.5, 79.8, 109.7, 113.60 (3C), 113.63 (2C), 113.7 (2C), 115.6, 117.0, 118.1, 118.4, 118.9, 123.4, 124.4, 125.6, 129.1, 129.7 (3C), 130.1, 131.1 (2C), 133.0 (2C), 133.2, 134.5, 134.7, 140.2, 158.1, 159.27, 159.32; HRMS (ESI$^+$) calcd for $C_{41}H_{40}BrN_2O_4$ (MH$^+$): 703.2163, found 703.2163.

{[7-(*tert*-Butoxy)-3-(4-methoxyphenethyl)-1,4-bis(4-methoxyphenyl)-3,6-dihydropyrrolo[2,3-*c*]carbazol-2-yl}-(4-methoxyphenyl)methanol (33) [41]. To a solution of **32** (266 mg, 0.378 mmol) in THF (10 mL) was added MeLi (1.17 M in Et_2O; 335 μL, 0.416 mmol) dropwise at −78 °C under Ar. After the mixture was stirred for 15 min, *n*-BuLi (1.64 M in *n*-hexane; 253 μL, 0.416 mmol) was added dropwise at −78 °C. Subsequently, *p*-methoxybenzaldehyde was added slowly at −78 °C. The mixture was warmed to room temperature and stirred for 3.5 h. The mixture was diluted with saturated aqueous NH_4Cl, and the aqueous layer was extracted twice with EtOAc. The combined organic layer was washed with H_2O and brine, dried over Na_2SO_4, filtered, and concentrated in vacuo. The residue was purified twice by column chromatography with (hexane/$CHCl_3$ = 3/1) and ($CHCl_3$/Et_2O = 9/1) to give **33** (214 mg, 74%) as an off-white solid: mp 239 °C (dec.); IR (neat) 3399 (NH); ^1H NMR (500 MHz, $CDCl_3$) δ: 1.46 (s, 9H), 2.03 (br s, 1H), 2.12 (d, $J = 3.5$ Hz, 1H), 2.45–2.51 (m, 1H), 3.71 (s, 3H), 3.78 (s, 3H), 3.88 (s, 3H), 3.92 (s, 3H), 3.96–4.04 (m, 2H), 5.92 (d, $J = 8.0$ Hz, 1H), 6.05 (d, $J = 3.5$ Hz, 1H), 6.18 (d, $J = 7.0$ Hz, 2H), 6.58 (d, $J = 8.5$ Hz, 2H), 6.67 (t, $J = 8.0$ Hz, 1H), 6.84 (d, $J = 8.5$ Hz, 2H), 6.92 (d, $J = 7.0$ Hz, 1H), 6.97 (dd, $J = 8.5, 2.5$ Hz, 1H), 7.02–7.06 (m, 3H), 7.23 (s, 1H), 7.30 (d, $J = 8.5$ Hz, 2H), 7.48–7.50 (m, 2H), 7.53 (dd, $J = 9.0, 2.5$ Hz, 1H), 7.59 (dd, $J = 8.0, 2.5$ Hz, 1H), 8.26 (s, 1H); ^{13}C NMR (125 MHz, $CDCl_3$) δ: 29.1 (3C), 35.5 47.6, 55.2, 55.3, 55.47, 55.51, 67.6, 79.8, 110.8, 113.3 (2C), 113.4, 113.5, 113.6 (2C), 113.8, 113.9, 114.3, 117.0, 118.1, 118.3, 119.0, 123.5, 124.6, 126.0, 126.8 (2C), 128.7, 129.6 (2C), 129.7, 130.6, 131.1, 131.3, 133.18, 133.21, 133.7, 134.2, 134.5, 134.8, 137.9, 140.1, 157.8, 158.6, 159.1, 159.3; HRMS (FAB) calcd for $C_{49}H_{49}N_2O_6$ (MH$^+$): 761.3585, found 761.3586.

{5-Bromo-7-(*tert*-butoxy)-3-(4-methoxyphenethyl)-1,4-bis(4-methoxyphenyl)-3,6-dihydropyrrolo[2,3-*c*]carbazol-2-yl}(4-methoxyphenyl)methanol (34). To a solution of **33** (15 mg, 0.019 mmol) in THF (1.0 mL) was added NBS (3.6 mg, 0.020 mmol) at −78 °C. The mixture was stirred at −78 °C under Ar for 0.5 h. The mixture was warmed to room temperature and stirred for 1 h. The mixture was diluted with brine, and the aqueous layer was extracted twice with EtOAc. The combined organic layer was washed with H_2O and brine, dried over Na_2SO_4, filtered, and concentrated in vacuo. The residue was purified by column chromatography (hexane/EtOAc = 2/1) to give **34** (8.0 mg, 50%) as an off-white solid: mp 137–142 °C; IR (neat) 3449 (NH); ^{1}H NMR (500 MHz, CDCl$_3$) δ: 1.48 (s, 9H), 2.15–2.20 (m, 2H), 2.59–2.65 (m, 1H), 3.67–3.82 (m, 8H), 3.87 (s, 3H), 3.90 (s, 3H), 5.87 (d, J = 8.5 Hz, 1H), 6.02 (d, J = 1.5 Hz, 1H), 6.39 (d, J = 8.0 Hz, 2H), 6.62–6.64 (m, 2H), 6.69 (t, J = 8.0 Hz, 1H), 6.79–6.81 (m, 2H), 6.94–6.99 (m, 2H), 7.02–7.07 (m, 3H), 7.22 (d, J = 8.5 Hz, 2H), 7.36 (dd, J = 8.0, 2.0 Hz, 1H), 7.43–7.53 (m, 3H), 8.49 (s, 1H); ^{13}C NMR (125 MHz, CDCl$_3$) δ: 29.1 (3C), 35.8, 47.0, 55.17, 55.25, 55.32, 55.5, 67.6, 80.1, 105.3, 113.3 (2C), 113.55, 113.65 (3C), 113.8, 113.9, 114.6, 117.6, 117.9, 118.9, 119.3, 122.5, 124.8. 125.1, 126.7 (2C), 129.3, 129.6 (2C), 129.9, 130.6, 131.2, 132.2, 132.3, 133.1, 133.2 (2C), 133.9, 134.2, 138.3, 140.4, 157.8, 158.6, 159.39, 159.41; HRMS (FAB) calcd for C$_{49}$H$_{48}$BrN$_2$O$_6$ (MH$^+$): 839.2690, found 839.2690.

{5-Bromo-7-(*tert*-butoxy)-3-(4-methoxyphenethyl)-1,4-bis(4-methoxyphenyl)-3,6-dihydropyrrolo[2,3-*c*]carbazol-2-yl}(4-methoxyphenyl)methanone (35) [41]. To a solution of **34** (8.0 mg, 9.5 μmol), NMO (2.2 mg, 0.019 mmol), and MS 4Å in CH$_2$Cl$_2$ (1.0 mL) was added TPAP (1.0 mg, 2.8 μmol) at 0 °C. After 1 h, the mixture was stirred at room temperature under Ar for 14 h. The mixture was filtered through a pad of Celite and concentrated in vacuo. The residue was purified by column chromatography (hexane/EtOAc = 2/1–3/2) to give **35** (8.0 mg, quant) as a yellow amorphous solid: mp 105–108 °C; IR (neat) 3595 (NH), 1687 (C=O); ^{1}H NMR (500 MHz, CDCl$_3$) δ: 1.49 (s, 9H), 2.55–2.59 (m, 2H), 3.70 (s, 3H), 3.79 (s, 3H), 3.81 (s, 3H), 3.85–3.91 (m, 5H), 5.95 (d, J = 6.4 Hz, 1H), 6.55–6.63 (m, 4H), 6.65–6.70 (m, 3H), 6.78–6.81 (m, 2H), 6.95 (d, J = 6.1 Hz, 1H), 7.07–7.10 (m, 2H), 7.28–7.30 (m, 2H), 7.47–7.50 (m, 2H), 7.57–7.62 (m, 2H), 8.56 (s, 1H); ^{13}C NMR (125 MHz, CDCl$_3$) δ: 29.1 (3C), 36.8, 47.4, 55.1, 55.36, 55.38, 55.4, 80.1, 107.2, 113.26 (2C), 113.29 (2C), 113.5 (2C), 113.9 (2C), 115.0, 117.5, 118.9, 119.8, 121.0, 121.9, 124.9, 125.0, 128.1, 129.5 (2C), 130.1, 131.0, 131.1, 131.2, 131.8 (2C), 132.3 (2C), 133.1 (2C), 133.5, 134.2, 136.6, 140.5, 158.0, 159.0, 159.6, 163.3, 190.1; HRMS (ESI$^+$) calcd for C$_{49}$H$_{46}$BrN$_2$O$_6$ (MH$^+$): 837.2534, found 837.2538.

{7-(*tert*-Butoxy)-5-methoxy-3-(4-methoxyphenethyl)-1,4-bis(4-methoxyphenyl)-3,6-dihydropyrrolo[2,3-*c*]carbazol-2-yl}(4-methoxyphenyl)methanone (12a) [40]: **formal synthesis of dictyodendrin E.** To a solution of **35** (14 mg, 0.017 mmol) and CuI (9.5 mg, 0.050 mmol) in DMF (1.0 mL) at room temperature was added NaOMe (4 M solution in MeOH; 71 μL), prepared separately in a loosely capped vial by the portionwise addition of sodium metal (92 mg, 4.0 mmol) to vigorously stirred MeOH (1.0 mL). The mixture was stirred at 80 °C in preheated bath under Ar for 1 h. The mixture was cooled to room temperature and diluted with saturated aqueous NH$_4$Cl, and the aqueous layer was extracted twice with EtOAc. The combined organic layer was washed with H$_2$O and brine, dried over Na$_2$SO$_4$, filtered, and concentrated in vacuo. The residue was purified by column chromatography (hexane/EtOAc = 2/1–3/2) to give **12a** (11 mg, 81%) as an yellow solid: mp 216–218 °C; IR (neat) 3350 (NH), 1618 (C=O); ^{1}H NMR (500 MHz, CDCl$_3$) δ: 1.48 (s, 9H), 2.52–2.55 (m, 2H), 3.64 (s, 3H), 3.69 (s, 3H), 3.78 (s, 3H), 3.80 (s, 3H), 3.92 (s, 3H), 4.00–4.02 (m, 2H), 5.94 (d, J =

8.0 Hz, 1H), 6.52 (d, J = 8.5 Hz, 2H), 6.60 (d, J = 8.0 Hz, 2H), 6.63–6.67 (m, 3H), 6.78 (d, J = 8.0 Hz, 2H), 6.93 (d, J = 8.0 Hz, 1H), 7.08 (d, J = 8.0 Hz, 2H), 7.28 (d, J = 8.5 Hz, 2H), 7.54 (d, J = 8.5 Hz, 2H), 7.60 (d, J = 9.0 Hz, 2H), 8.43 (s, 1H); ^{13}C NMR (125 MHz, CDCl$_3$) δ: 29.1 (3C), 36.6, 47.5, 55.1, 55.4 (3C), 61.1, 79.9, 113.1 (2C), 113.2 (2C), 113.5 (2C), 113.8 (2C), 115.8, 117.3, 118.0, 118.6, 119.49, 119.54, 122.1, 125.0, 127.2, 128.6, 129.47, 129.55 (2C), 130.3 131.1, 131.7, 132.20 (2C), 132.22 (2C), 133.1 (2C), 134.4, 136.3, 140.4, 142.6, 158.0, 158.9, 159.4, 163.0, 189.8; HRMS (ESI$^+$) calcd for C$_{50}$H$_{49}$N$_2$O$_7$ (MH$^+$): 789.3534, found 789.3540.

{7-Hydroxy-5-methoxy-3-(4-methoxyphenethyl)-1,4-bis(4-methoxyphenyl)-3,6-dihydropyrrolo[2,3-c]carbazol-2-yl}(4-methoxyphenyl)methanone   (S8) [19]. To a solution of **12a** (11 mg, 0.014 mmol) and pentamethylbenzene (6.1 mg, 0.041 mmol) in CH$_2$Cl$_2$ (1.0 mL) was added BCl$_3$ (1.0 M n-heptane, 34 μL, 0.034 mmol) at −78 °C. The mixture was stirred at −78 °C under Ar for 10 min. The mixture was diluted with 10% MeOH/CHCl$_3$ at −78 °C and concentrated in vacuo. The residue was purified by column chromatography (hexane/EtOAc = 3/2–1/1) to give **S8** (9.8 mg, 99%) as a yellow amorphous solid: ^{1}H NMR (500 MHz, CD$_2$Cl$_2$) δ: 2.50–2.53 (m, 2H), 3.63 (s, 3H), 3.68 (s, 3H), 3.78 (s, 3H), 3.79 (s, 3H), 3.89 (s, 3H), 3.92–3.96 (m, 2H), 5.72–5.86 (br s, 2H), 6.50–6.53 (m, 2H), 6.56–6.71 (m, 6H), 6.76–6.79 (m, 2H), 7.05–7.08 (m, 2H), 7.23–7.26 (m, 2H), 7.53–7.59 (m, 4H), 8.62–8.96 (br s, 1H); ^{13}C NMR (125 MHz, CD$_2$Cl$_2$) δ: 36.3, 47.3, 54.8, 55.09, 55.14, 55.2, 60.9, 108.7, 112.9 (2C), 113.2 (2C), 113.5 (2C), 115.2, 117.0, 118.1, 118.6, 119.1, 121.7, 124.9, 126.9, 128.3, 128.5, 129.2 (2C), 129.7, 130.1, 131.0, 131.3, 131.96 (2C), 132.04 (2C), 132.8 (2C), 135.9, 140.9, 142.6, 157.9, 158.8, 159.3, 163.0. 189.5.

5-Methoxy-2-(4-methoxybenzoyl)-3-(4-methoxyphenethyl)-1,4-bis(4-methoxyphenyl)-3,6-dihydropyrrolo[2,3-c]carbazol-7-yl (2,2,2-trichloroethyl) sulfate (S9) [19]. To a solution of **S8** (9.4 mg, 0.013 mmol) and DABCO (3.0 μL, 0.027 mmol) in CH$_2$Cl$_2$ (1.0 mL) was added 2,2,2-trichloroethyl chlorosulfate [10] (6.6 μL, 0.027 mmol) at room temperature. The mixture was stirred at room temperature under Ar for 1 h. The mixture was diluted with saturated aqueous NH$_4$Cl, and the aqueous layer was extracted twice with EtOAc. The combined organic layer was washed with H$_2$O and brine, dried over Na$_2$SO$_4$, filtered, and concentrated in vacuo. The residue was purified by column chromatography (hexane/EtOAc = 1/1) to give **S9** (10 mg, 85%) as a yellow amorphous solid: ^{1}H NMR (500 MHz, CD$_2$Cl$_2$) δ: 2.51–2.53 (m, 2H), 3.64 (s, 3H), 3.69 (s, 3H), 3.78 (s, 3H), 3.81 (s, 3H), 3.91 (s, 3H), 3.93–3.96 (m, 2H), 4.85 (s, 2H), 5.32 (d, J = 1.5 Hz, 1H), 6.17 (d, J = 8.0 Hz, 1H), 6.52 (d, J = 9.0 Hz, 2H), 6.60 (d, J = 8.5 Hz, 1H), 6.71 (d, J = 9.0 Hz, 2H), 6.77 (t, J = 9.0 Hz, 1H), 6.81 (d, J = 8.5 Hz, 2H), 7.10 (d, J = 9.0 Hz, 2H), 7.26 (d, J = 9.0 Hz, 2H), 7.31 (d, J = 7.5 Hz, 1H), 7.56 (d, J = 8.5 Hz, 2H), 7.58 (d, J = 9.0 Hz, 2H), 8.67 (s, 1H); ^{13}C NMR (125 MHz, CD$_2$Cl$_2$) δ: 36.9, 47.9, 55.4, 55.71, 55.76, 55.79, 61.7, 81.2, 92.7, 113.5, 113.6, 113.8, 114.2, 115.3, 116.3, 119.0, 119.5, 119.9, 121.6, 124.6, 127.1, 127.5, 128.6, 129.8, 130.5, 130.9, 131.0, 131.67, 131.71, 132.5, 132.6, 133.4, 135.1, 136.9, 142.8, 158.6, 159.5, 160.0, 163.7, 189.8.

Dictyodendrin B (2) [19]. To a solution of **S9** (9.8 mg, 0.010 mmol) and *n*-Bu$_4$NI (96 mg, 0.26 mmol) in CH$_2$Cl$_2$ (1.0 mL) was added BCl$_3$ (1.0 M *n*-heptane;

34 μL, 0.034 mmol) at 0 °C. The mixture was stirred at room temperature under Ar for 0.5 h. The mixture was diluted with saturated aqueous NH_4Cl, and the aqueous layer was extracted twice with EtOAc. The combined organic layer was washed with H_2O and brine, dried over Na_2SO_4, filtered, and concentrated in vacuo. The residue was purified by diol silica column chromatography (CH_2Cl_2/MeOH = 10/1) to give crude **S10** as a yellow amorphous solid. Crude **S10**, zinc dust (1.8 mg, 0.028 mmol), and HCO_2NH_4 (2.6 mg, 0.041 mmol) were dissolved with MeOH (1 mL) at room temperature. The mixture was stirred under Ar at room temperature for 2 h and 50 °C for 2 h. The mixture was filtered through a pad of Celite. The filtrate was concentrated in vacuo. The residue was purified by diol silica column chromatography (CH_2Cl_2/MeOH = 3/1) to give dictyodendrin B (**2**) (3.2 mg, 29% in 2 steps) as a yellow amorphous solid: IR (neat) 3629 (NH), 1607 (C=O); ^{1}H NMR (500 MHz, CD_3OD) δ: 2.46–2.49 (m, 2H), 3.94–3.97 (m, 2H), 6.01 (d, J = 8.0 Hz, 1H), 6.41 (d, J = 8.5 Hz, 2H), 6.47 (d, J = 8.0 Hz, 2H), 6.56–6.60 (m, 3H), 6.65 (d, J = 8.0 Hz, 2H), 7.02–7.06 (m, 4H), 7.18 (d, J = 8.0 Hz), 7.33 (d, J = 9.0 Hz, 2H), 7.45 (d, J = 8.0 Hz, 2H), 8.55 (s, 1H); ^{13}C NMR (125 MHz, CD_3OD) δ: 37.7, 48.5, 112.6, 115.5 (2C), 115.6 (2C), 115.95 (2C), 116.04, 116.7 (2C), 117.2, 118.2, 118.8, 122.7, 125.7, 126.8, 127.1, 129.2, 129.3, 130.6 (2C), 131.8, 133.8, 133.97 (2C), 134.01 (2C), 134.26 (2C), 134.31, 136.2, 138.7, 141.7, 156.8, 157.7, 158.7, 163.3, 191.9; HRMS (ESI$^-$) calcd for $C_{41}H_{29}N_2O_{10}S$ ([M]$^-$): 741.1548, found 741.1552.

Methyl   2-{7-(*tert*-butoxy)-5-methoxy-3-(4-methoxyphenethyl)-1,4-bis(4-methoxyphenyl)-3,6-dihydropyrrolo[2,3-*c*]carbazol-2-yl}-2-oxoacetate (36). To a solution of **13a** (39 mg, 0.059 mmol) in THF (3.0 mL) was added (COCl)$_2$ (51 μL, 0.59 mmol) at room temperature. The mixture was stirred at 55 °C under Ar for 4 h. The mixture was diluted with MeOH and neutralized with aqueous NaHCO$_3$, and the aqueous layer was extracted twice with EtOAc. The combined organic layer was washed with H_2O and brine, dried over Na_2SO_4, filtered, and concentrated in vacuo. The residue was purified by column chromatography (hexane/EtOAc = 3/1) to give **36** (38 mg, 87%) as an yellow solid: mp 155–160 °C; IR (neat) 3379 (NH), 1631 (C=O); ^{1}H NMR (500 MHz, CDCl$_3$) δ: 1.46 (s, 9H), 2.62–2.65 (m, 2H), 3.31 (s, 3H), 3.62 (s, 3H), 3.74 (s, 3H), 3.93 (s, 3H), 3.96 (s, 3H), 4.36–4.39 (m, 2H), 5.52 (d, J = 8.0 Hz, 1H), 6.64–6.69 (m, 3H), 6.76–6.78 (d, J = 7.0 Hz, 2H), 6.93 (d, J = 7.0 Hz, 1H), 7.06–7.10 (m, 4H), 7.46–7.52 (m, 4H), 8.48 (s, 1H); ^{13}C NMR (125 MHz, CDCl$_3$) δ: 29.1(3C), 36.5, 47.8, 52.1, 55.2, 55.4, 55.6, 61.3, 80.1, 113.3 (2C), 113.5 (2C), 114.0 (2C), 115.8, 117.3, 117.7, 118.7, 119.1, 120.3, 124.7, 126.59, 126.63, 129.2, 129.5, 129.8 (2C), 130.3, 130.4, 132.1 (2C), 134.15, 134.18 (2C), 134.7, 140.7, 146.0, 158.0, 159.6, 160.1, 164.7, 180.0; HRMS (ESI$^+$) calcd for $C_{45}H_{45}N_2O_8$ (MH$^+$): 741.3170, found 741.3168.

tert-Butyl 2-{7-(*tert*-butoxy)-5-methoxy-3-(4-methoxyphenethyl)-1,4-bis(4-methoxyphenyl)-3,6-dihydropyrrolo[2,3-*c*]carbazol-2-yl}-2-oxoacetate (37). To a solution of **13a** (38.0 mg, 0.0513 mmol) in THF (2.0 mL) was added *t*-BuOLi (33.3 mg, 0.416 mmol) at 0 °C. The mixture was stirred at room temperature for 10 h. The mixture was quenched with aqueous NH₄Cl, and the whole was extracted twice with EtOAc. The combined organic layer was washed with H₂O and brine, dried over Na₂SO₄, filtered, and concentrated in vacuo. The residue was purified by column chromatography (hexane/EtOAc = 3/1) to give **37** (25.0 mg, 61%) as an yellow solid: mp 190–193 °C; IR (neat); 3566 (NH), 1642 (C=O); ^{1}H NMR (500 MHz, CDCl₃) δ: 1.24 (s, 9H), 1.46 (s, 9H), 2.63–2.66 (m, 2H), 3.60 (s, 3H), 3.73 (s, 3H), 3.92 (s, 3H), 3.95 (s, 3H), 4.29–4.32 (m, 2H), 5.49 (d, J = 8.0 Hz, 1H), 6.64–6.69 (m, 3H), 6.77 (d, J = 9.0 Hz, 2H), 6.93 (d, J = 8.0 Hz, 1H), 7.06–7.09 (m, 4H), 7.49–7.51 (m, 4H), 8.45 (s, 1H); ^{13}C NMR (125 MHz, CDCl₃) δ: 27.7(3C), 29.1(3C), 36.5, 47.5, 55.2, 55.4, 55.6, 61.2, 80.0, 83.4, 113.4 (2C), 113.8 (2C), 114.0 (2C), 115.8, 117.3, 117.7, 118.88, 118.91, 120.3, 124.8, 126.8, 127.1, 129.1, 129.5, 129.78 (2C), 129.80, 130.5, 132.01 (2C), 134.18, 134.20 (2C), 134.24, 140.6, 145.3, 158.0, 159.5, 160.3, 163.5, 181.1; HRMS (ESI⁺) calcd for C₄₈H₅₁N₂O₈ (MH⁺): 783.3640, found 783.3642.

tert-Butyl 2-{7-(*tert*-butoxy)-5-methoxy-3-(4-methoxyphenethyl)-1,4-bis(4-methoxyphenyl)-3,6-dihydropyrrolo[2,3-*c*]carbazol-2-yl}-2-hydroxy-2-(4-methoxyphenyl)acetate (38). To a solution of **37** (25.0 mg, 0.0319 mmol) in THF (1.0 mL) was added 4-methoxyphenylmagnesium bromide (0.5 M THF; 640 μL, 0.319 mmol) at −40 °C. The mixture was stirred at −40 °C under Ar for 1 h. The mixture was quenched with aqueous NH₄Cl, and the whole was extracted twice with EtOAc. The combined organic layer was washed with H₂O and brine, dried over Na₂SO₄, filtered, and concentrated in vacuo. The residue was purified by column chromatography (hexane/EtOAc = 3/1) to give **38** (23.9 mg, 84%) as an off-white solid: mp 123–127 °C; IR (neat) 3481 (NH); ^{1}H NMR (500 MHz, CDCl₃) δ: 0.960 (s, 9H), 1.45 (s, 9H), 2.21–2.26 (m, 1H), 2.38–2.43 (m, 1H), 3.27–3.36 (br s, 1H), 3.55 (s, 3H), 3.60–3.67 (m, 1H), 3.70 (s, 3H), 3.77 (s, 3H), 3.86 (s, 3H), 3.91 (s, 3H), 4.69 (s, 1H), 5.50 (d, J = 8.0 Hz, 1H), 6.58–6.64 (m, 3H), 6.68–6.71 (m, 4H), 6.89–6.94 (m, 4H), 7.05 (br s, 1H), 7.41–7.42 (m, 4H), 7.53–7.54 (br s, 1H), 7.67–7.68 (br s, 1H), 8.34 (s, 1H); ^{13}C NMR (125 MHz, CDCl₃) δ: 27.2(3C), 29.1(3C), 34.4, 48.2, 55.13, 55.15, 55.3, 55.6, 60.6, 77.6, 79.8, 83.8, 112.8(2C), 112.9(2C), 113.3(2C), 113.6(2C), 115.6, 116.0, 117.2, 118.0, 118.1, 119.5, 119.8, 122.3, 125.1, 127.5, 128.2(2C), 129.5, 129.6(2C), 129.7, 131.2, 131.4, 134.0(2C), 134.5, 134.9(2C), 136.5, 140.2, 140.3, 157.6, 158.9, 159.0, 159.3, 173.2; HRMS (ESI⁺) calcd for C₅₅H₅₉N₂O₄₉ (MH⁺): 891.4215, found 891.4217.

Methyl 2-{7-(*tert*-butoxy)-5-methoxy-3-(4-methoxyphenethyl)-1,4-bis(4-methoxyphenyl)-3,6-dihydropyrrolo[2,3-*c*]carbazol-2-yl}-2-hydroxy-2-(4-methoxyphenyl)acetate (S12). To a solution of **36** (7.7 mg, 0.010 mmol) in THF (0.20 mL) was added 5 M NaOH aq. (0.2 mL) at room temperature. The mixture was stirred at room temperature for 1 h. The mixture was diluted with saturated aqueous NH$_4$Cl, and the aqueous layer was extracted twice with EtOAc. The combined organic layer was washed with H$_2$O and brine, dried over Na$_2$SO$_4$, filtered, and concentrated in vacuo. The residue was filtered through by short column chromatography (CHCl$_3$/MeOH = 9/1) to give crude **S11**. To a solution of the resulting crude **S11** in THF was added 4-methoxyphenylmagnesium bromide (0.5 M THF; 0.21 mL, 0.11 mmol) at room temperature. The mixture was stirred at room temperature for 15 min. The mixture was diluted with aqueous NH$_4$Cl, and the aqueous layer was extracted twice with EtOAc. The combined organic layer was washed with H$_2$O and brine, dried over Na$_2$SO$_4$, filtered, and concentrated in vacuo to give a resulting crude **39**, which was used for the next step without further purification. To a solution of the crude **39** in THF/MeOH (1/1, 0.3 mL) was added trimethylsilyldiazomethane (2 M solution in THF; 50 μL, 0.10 mmol) at room temperature. The mixture was stirred at room temperature for 5 min and concentrated in vacuo. The residue was purified by column chromatography (hexane/EtOAc = 3/1) to give **S12** (6.0 mg, 68%) as an off-white solid: mp 145–150 °C; IR (neat) 3476 (NH), 1729 (C=O); ^{1}H NMR (600 MHz, CDCl$_3$) δ: 1.45 (s, 9H), 2.14 (br s, 1H), 2.49 (ddd, J = 12.6, 12.6, 4.8 Hz, 1H), 2.97 (s, 3H), 3.35–3.43 (br s, 1H), 3.53 (s, 3H), 3.64–3.71 (m, 4H), 3.78 (s, 3H), 3.86 (s, 3H), 3.95 (s, 3H), 4.55 (br s, 1H), 5.33 (d, J = 8.4 Hz, 1H), 6.55–6.62 (m, 5H), 6.79 (d, J = 9.6 Hz, 2H), 6.90–6.92 (m, 2H), 7.03–7.08 (m, 3H), 7.42 (d, J = 8.4 Hz, 3H), 7.50–7.51 (m, 2H), 7.69 (d, J = 7.8 Hz, 1H), 8.37 (s, 1H); ^{13}C NMR (150 MHz, CDCl$_3$) δ: 29.1(3C), 34.7, 47.9, 53.4, 55.15, 55.18, 55.3, 55.7, 60.7, 79.9, 112.8, 113.2 (2C), 113.3 (2C), 113.5, 113.6, 114.2, 115.5, 117.3 (2C), 117.9, 118.2, 119.2, 121.8, 125.0, 127.3, 128.1 (2C), 129.0, 129.4, 129.6 (2C), 130.3, 131.3, 132.0, 132.3, 134.5, 134.8 (2C), 135.1, 135.7, 140.2, 140.7, 157.6, 159.0, 159.1, 159.4, 174.1; HRMS (ESI$^+$) calcd for C$_{52}$H$_{53}$N$_2$O$_9$ (MH$^+$): 849.3746, found 849.3741.

Methyl 2-{7-(*tert*-butoxy)-5-methoxy-3-(4-methoxyphenethyl)-1,4-bis(4-methoxyphenyl)-3,6-dihydropyrrolo[2,3-*c*]carbazol-2-yl}-2-(4-methoxyphenyl)acetate (11). A mixture of **S12** (2.7 mg, 0.0032 mmol) and Pd(OH)$_2$/C (*ca.* 50 wt% on carbon, 2.0 mg) in *i*-PrOH (0.20 mL) was stirred under a hydrogen atmosphere at room temperature for 10 h. The resulting suspension was filtered through a Celite pad, and the pad was washed with EtOAc. The filtrate was concentrated *in vacuo*. The residue was purified by flash chromatography on silica gel (hexane/EtOAc = 2/1) to afford **11** (1.4 mg, 53%) as a white solid: mp 100–105 °C; IR (neat) 3594 (NH), 1723 (C=O); ^{1}H NMR (500 MHz, CDCl$_3$) δ: 1.45 (s, 9H), 2.22–2.23 (m, 2H), 3.53 (s, 3H), 3.61 (s, 3H), 3.71 (s, 3H), 3.77 (s, 3H), 3.83–3.89 (m, 2H), 3.90 (s, 3H), 3.92 (s, 3H), 5.34 (s, 1H), 5.80 (d, J = 7.2 Hz, 1H), 6.26 (d, J = 9.0 Hz, 2H), 6.58–6.60 (m, 2H), 6.65 (t, J = 8.4 Hz, 1H), 6.79–6.81 (m, 2H), 6.92 (d, J = 8.4 Hz, 1H), 7.00–7.08 (m, 4H), 7.14–7.15 (m, 2H), 7.42–7.44 (m, 1H), 7.55–7.62 (m, 2H), 8.34 (s, 1H); ^{13}C NMR (125 MHz, CDCl$_3$) δ: 29.1 (3C), 35.4, 47.2, 47.3, 52.3, 55.2, 55.3, 55.4, 55.5, 60.8, 80.0, 113.3 (2C), 113.5, 113.6 (2C), 113.77, 113.81 (2C), 115.2, 117.5, 117.8, 118.0, 118.5 (2C), 119.1, 125.2, 128.4, 128.9, 129.0 (2C), 129.4(2C), 129.5, 130.1, 132.68 (2C), 132.74 (2C), 133.2, 133.5, 133.7, 134.6, 140.2, 140.4, 157.8, 158.5, 159.25, 159.31, 171.9; HRMS (ESI$^+$) calcd for C$_{52}$H$_{52}$N$_2$O$_8$ (MH$^+$): 833.3796, found 833.3797.

4-(4-Methoxyphenyl)-3,6-dihydropyrrolo[2,3-c]carbazol-7-ol (40). To a solution of **24** (38 mg, 0.098 mmol) and pentamethylbenzene (48 mg, 0.32 mmol) in CH_2Cl_2 (1.0 mL) was added BCl_3 (1.0 M in n-heptane; 0.13 mL, 0.13 mmol) at $-78\,°C$. The mixture was stirred at $-78\,°C$ under Ar for 1 h. The mixture was quenched with 10% $MeOH/CHCl_3$ at $-78\,°C$ and concentrated in vacuo. The residue was purified by column chromatography (hexane/EtOAc = 3/2–1/1) to give **40** (3.1 mg, 10%) as a brown solid: mp 108–110 °C; IR (neat) 3417 (NH); ^1H NMR (500 MHz, $CDCl_3$) δ: 3.91 (s, 3H), 5.08 (br s, 1H), 6.83 (d, $J = 7.2$ Hz, 1H), 7.07–7.16 (m, 4H), 7.32 (s, 1H), 7.38–7.40 (dd, $J = 2.4, 2.4$ Hz, 1H), 7.63–7.67 (m, 2H), 7.88 (d, $J = 8.0$ Hz, 1H), 8.34 (s, 1H), 8.56 (s, 1H); ^{13}C NMR (125 MHz, $CDCl_3$) δ: 55.4, 101.4, 106.3, 109.1, 113.8, 114.2, 114.6(2C), 119.4, 121.3, 124.0, 125.0, 125.6, 128.5, 129.0, 129.5(2C), 132.0, 134.9, 140.9, 159.1; HRMS (ESI$^+$) calcd for $C_{21}H_{17}N_2O_2$ (MH$^+$): 329.1285, found 329.1285.

10cd　　　　　　　　　**S14**　　　　　　　　　**41**

4-(4-Methoxyphenyl)-1,6-dihydropyrrolo[3,2-c]carbazol-7-ol (41). To a solution of **10 cd** (15 mg, 0.029 mmol) in THF (0.3 mL) was added 5 M NaOMe (30 μL, 0.015 mmol) at room temperature. The mixture was stirred at room temperature under Ar for 1 h. To the mixture was added saturated aqueous NH_4Cl, and the aqueous layer was extracted twice with EtOAc. The combined organic layer was washed with H_2O and brine, dried over Na_2SO_4, filtered, and concentrated in vacuo to give crude **S14**. A mixture of **S14** and $Pd(OH)_2/C$ (*ca.* 50 wt% on carbon, 7.8 mg) in EtOH (1.0 mL) was stirred under a hydrogen atmosphere at room temperature for 22 h. The resulting suspension was filtered through a Celite pad, and the pad was washed with EtOAc. The filtrate was concentrated *in vacuo*. The residue was purified by flash chromatography on silica gel (hexane/EtOAc = 3/1) to afford **41** (5.5 mg, 92%) as an off-white solid: mp 115–118 °C; IR (neat) 3423 (NH); ^1H NMR (500 MHz, $CDCl_3$) δ: 3.91 (s, 3H), 5.19 (br s, 1H), 6.83–6.85 (m, 2H), 7.06–7.07 (m, 2H), 7.15 (t, $J = 8.4$ Hz, 1H), 7.30–7.32 (m, 2H), 6.67 (d, $J = 8.4$ Hz, 1H), 7.71–7.73 (m, 2H), 8.44 (s, 1H), 8.76 (s, 1H); ^{13}C NMR (125 MHz, $CDCl_3$) δ: 55.4, 103.4, 104.6, 106.9, 109.2, 113.2, 114.0 (2C), 119.8, 120.1, 121.5, 123.7, 128.1, 130.1 (2C), 130.2, 133.5, 134.2, 137.2, 141.1, 158.8; HRMS (ESI$^+$) calcd for $C_{21}H_{16}N_2O_2$ (MH$^+$): 329.1285, found 329.1280.

2.1.5 Preparation of Catalyst

[BrettPhosAu·MeCN]SbF$_6$ [42].

This catalyst was prepared according to the literature procedure for the synthesis of [LAu·MeCN]SbF$_6$ (L = {2-[2,4,6-(i-Pr)$_3$C$_6$H$_2$]C$_6$H$_4$}P(t-Bu)$_2$).[9] AgSbF$_6$ (0.6 M solution in CH$_2$Cl$_2$; 1.98 mL, 1.19 mmol) was added to a stirred solution of chloro[2-(dicyclohexylphosphino)-3,6-dimethoxy-2′,4′,6′-triisopropyl-1,1′-biphenyl]gold(I) (BrettPhosAuCl) (897 mg, 1.17 mmol) in MeCN (7.50 mL) and CH$_2$Cl$_2$ (7.50 mL), and the mixture was stirred at room temperature in the dark (using aluminum foil) for 8 h. The mixture was filtered through a pad of Celite, and the solvent was removed in vacuo to afford a white powder (1.29 g, quant): ^{1}H NMR (500 MHz, CDCl$_3$) δ: 0.90 (d, J = 6.3 Hz, 6H), 1.07–1.09 (m, 2H), 1.17–1.24 (m, 4H), 1.27 (d, J = 6.9 Hz, 6H), 1.33 (d, J = 6.9 Hz, 6H), 1.37–1.40 (m, 4H), 1.49–1.50 (m, 2H), 1.67–1.98 (m, 8H), 2.25–2.30 (m, 2H), 2.37 (s, 3H), 2.55–2.59 (m, 2H), 2.92–2.98 (m, 1H), 3.56 (s, 3H), 3.94 (s, 3H), 6.95–7.12 (m, 4H); ^{13}C NMR (125 MHz, CDCl$_3$) δ: 2.52, 24.1 (d, J = 19.2 Hz, 4C), 24.8 (2C), 25.6 (2C), 26.5 (d, J = 16.8 Hz, 2C), 27.0 (d, J = 13.2 Hz, 2C), 30.0 (2C), 30.6 (2C), 33.7, 34.6, 34.7, 38.1 (d, J = 36.0 Hz, 2C), 54.9, 56.0, 110.8 (d, J = 7.2 Hz), 114.2, 114.7, 115.1, 118.9, 121.6 (2C), 131.5 (d, J = 8.4 Hz), 136.7 (d, J = 13.2 Hz), 147.3, 149.2, 153.1 (d, J = 10.8 Hz), 154.8; HRMS (FAB) calcd for C$_{35}$H$_{53}$AuO$_2$P$^+$ (M–MeCN–SbF$_6$)$^+$: 733.3443, found 733.3444.

2.1.6 Biological Evaluations and Binding Mode Analysis

Growth Inhibition Assay. HCT116 cells were cultured in McCoy's 5A medium (GIBCO) supplemented with 10% (v/v) fetal bovine serum at 37 °C in a 5% CO$_2$ incubator. Growth inhibition assays using HCT116 cells were performed in 96-well plates (BD Falcon). HCT116 cells were seeded at 5000 cells/well in 50 µL of culture media, respectively, and were cultured for 6 h. 30 mM chemical compounds in DMSO were diluted 250-fold with the culture medium in advance. Following the addition of 40 µL of the fresh culture medium to the cell cultures, 30 µL of the chemical diluents was also added. The final volume of DMSO in the medium was equal to 0.1% (v/v). The cells under chemical treatment were incubated for further 72 h. The wells in the plates were washed twice with the cultured medium without phenol red.

After 1 h incubation with 100 μL of the medium, the cell culture in each well was supplemented with 20 μL of the MTS reagent (Promega), followed by incubation for additional 40 min. Absorbance at 490 nm of each well was measured using a Wallac 1420 ARVO SX multilabel counter (PerkinElmer).

CDK2/CycA2 and GSK3β kinase assay. CDK2/CycA2 and GSK3β inhibitory activities were evaluated by the off-chip mobility shift assay by the QuickScout® service from Carna Biosciences (Kobe, Japan). Human GST-fusion CDK2/CycA2 (1-298) with GST-CyclinA2 was co-expressed using baculovirus expression system. Human GST-GSK3β (1-420) was expressed using baculovirus expression system. GST-CDK2/CycA2 and GSK3β were purified by using glutathione Sepharose chromatography. Each chemical in DMSO at different concentrations was diluted fourfold with reaction buffer [20 mM HEPES (pH 7.5), 0.01% Triton X-100, 2 mM DTT]. For CDK2/CycA2 reactions, a combination of the compound, 1 μM modified histone H1, 5 mM $MgCl_2$, 27 μM ATP in reaction buffer (20 μL) was incubated with each CDK/2CycA2 in PP 384-well plates at room temperature for 1 h ($n = 2$). For GSK3β reactions, a combination of the compound, 1 μM CREBtide-p, 5 mM $MgCl_2$, 9.1 μM ATP in reaction buffer (20 μL) was incubated with each GSK3βin PP 384-well plates at room temperature for 1 h ($n = 2$). The reaction was terminated by addition of 70 μL of termination buffer (Carna Biosciences). Substrate and product were separated by electrophoretic means using the LabChip3000 system. The kinase reaction was evaluated by the product ratio, which was calculated from the peak heights of the substrate (S) and product (P): [P/(P + S)]. Inhibition data were calculated by comparing with no-enzyme controls for 100% inhibition and no-inhibitor reactions for 0% inhibition. IC_{50} values were calculated using GraphPad Prism 5 software (GraphPad Software Incorporated, La Jolla, CA, USA).

Screening for nuclear localization of the JEV core protein. The Huh7 cells stably expressing the sfGFP-JEV core (1 × 104) were cultured in a 96-well microplate (#655866; Greiner Bio-One) at 37 °C for 24 h, followed by incubation with compounds (final concentration 10 μM) for 24 h. Cells were fixed with 3.7% formaldehyde in PBS for 15 min at room temperature and then incubated with PBS containing Hoechst 33342 (Dojindo, final concentration 0.2 μg/mL) for nuclear staining. Fluorescence images were captured by a high-content imaging system Cell Voyager 7000S (CV7000S, Yokogawa Electric Corporation, Tokyo). The outline of the nucleus stained with Hoechst 33342 was automatically recognized, and the mean intensity of sfGFP expression area in the nucleus was quantitatively measured to obtain the average fluorescence intensity in each well.

References

1. Zhang W, Ready JM (2016) J Am Chem Soc 138:10684–10692
2. Zhang H, Conte MM, Khalil Z, Huang X-C, Capon RJ (2012) RSC Adv. 2:4209–4214
3. Gorin DJ, Toste FD (2007) Nature 446:395–403
4. Arcadi A (2008) Chem Rev 108:3266–3325
5. Shapiro ND, Toste FD (2010) Synlett 5:675–691

6. Wang Y, Muratore ME, Echavarren AM (2015) Chem Eur J 21:7332–7339
7. Qiana D, Zhang J (2015) Chem Soc Rev 44:677–698
8. Harris RJ, Widenhoefer RA (2016) Chem Soc Rev 45:4533–4551
9. Wetzel A, Gagosz F (2011) Angew Chem Int Ed 50:7354–7358
10. Lu B, Luo Y, Liu L, Ye L, Wang Y, Zhang L (2011) Angew Chem Int Ed 50:8358–8362
11. Asiri AM, Hashmi ASK (2016) Chem Soc Rev 45:4471–4503
12. Matsuda Y, Naoe S, Oishi S, Fujii N, Ohno H (2015) Chem Eur J 21:1463–1467
13. Tokimizu Y, Oishi S, Fujii N, Ohno H (2015) Angew Chem Int Ed 54:7862–7866
14. Taguchi M, Tokimizu Y, Oishi S, Fujii N, Ohno H (2015) Org Lett 17:6250–6253
15. Naoe S, Yoshida Y, Oishi S, Fujii N, Ohno H (2016) J Org Chem 81:5690–5698
16. Matsuoka J, Kumagai H, Inuki S, Oishi S, Ohno H (2019) J Org Chem 84:9358–9363
17. Kawada Y, Ohmura S, Kobayashi M, Nojo W, Kondo M, Matsuda Y, Matsuoka J, Inuki S, Oishi S, Wang C, Saito T, Uchiyama M, Suzuki T, Ohno H (2018) Chem Sci 9:8416–8425
18. Matsuoka J, Matsuda Y, Kawada Y, Oishi S, Ohno H (2017) Angew Chem Int Ed 56:7444–7448
19. Tokuyama H, Okano K, Fujiwara H, Noji T, Fukuyama T (2011) Chem Asian J 6:560–572
20. Pitts AK, O'Hara F, Snell RH, Gaunt MJ (2015) Angew Chem Int Ed 54:5451–5455
21. Fürstner A, Domostoj MM, Scheiper B (2006) J Am Chem Soc 128:8087–8094
22. Padwa A, Austin DJ, Gareau Y, Kassir JM, Xu SL (1993) J Am Chem Soc 115:2637–2647
23. Sindhu KS, Thankachan AP, Sajitha PS, Anilkumar G (2015) Org Biomol Chem 13:6891–6905
24. Marino JP, Nguyen HN (2002) J Org Chem 67:6841–6844
25. Quan Y, Qiu Z, Xie Z (2014) J Am Chem Soc 136:7599–7602
26. Chen XY, Wang L, Frings M, Bolm C (2014) Org Lett 16:3796–3799
27. Izgu EC, Hoye TR (2012) Tetrahedron Lett 53:4938–4941
28. Barral K, Moorhouse AD, Moses JE (2007) Org Lett 9:1809–1811
29. Harrisson P, Morris J, Marder TB, Steel PG (2009) Org Lett 11:3586–3589
30. Maeda C, Todaka T, Ema T (2015) Org Lett 17:3090–3093
31. Katritzky AR, Rewcastle GW (1988) Vazquez de Miguel, L. M. J Org Chem 53:794–799
32. Hartung CG, Fecher A, Chapell B, Snieckus V (2003) Org Lett 5:1899–1902
33. Chelucci G, Baldino S, Ruiu A (2012) J Org Chem 77:9921–9925
34. a Mori Y, Okabayashi T, Yamashita T, Zhao Z, Wakita T, Yasui K, Hasebe F, Tadano M, Konishi E, Moriishi K, Matsuura Y (2005) J Virol 79:3448–3458; b Tokunaga M, Miyamoto Y, Suzuki T, Otani M, Inuki S, Esaki T, Nagao C, Mizuguchi K, Ohno H, Yoneda Y, Okamoto T, Oka M, Matsuura Y (2020) Virology 541:41–51
35. Naoe S, Suzuki Y, Hirano K, Inaba Y, Oishi S, Fujii N, Ohno H (2012) J Org Chem 77:4907–4916
36. a Liu G, Xu G, Li J, Ding D, Sun J (2014) Org Biomol Chem 12:1387–1390; b Sakai N, Annaka K, Konakahara T (2006) J Org Chem 71:3653–3655
37. Nie X, Wang G (2006) J Org Chem 71:4734–4741
38. a Padwa A, Austin DJ, Gareau Y, Kassir JM, Xu SL (1993) J Am Chem Soc 115:2637–2647; b Fiandanese V, Bottalico D, Marchese G, Punzi A (2008) Tetrahedron 64:7301–7306
39. Kueh JTB, Choi KW, Brimble MA (2012) Org Biomol Chem 10:5993–6002
40. a Aalten HL, van Koten G, Grove DM, Kuilman T, Piekstra OG, Hulshof LA, Sheldon RA (1989) Tetrahedron 45:5565–5578; b Kikugawa Y, Aoki Y, Sakamoto T (2001) J Org Chem 66:8612–8615; c Bhosale SM, Momin AA, Kusurkar RS (2012) Tetrahedron 68; 6420–6426; d Miyamoto H, Hirano T, Okawa Y, Nakazaki A, Kobayashi S (2013) Tetrahedron 69:9481–9493, e Pitts AK, O'Hara F, Snell RH, Gaunt MJ (2015) Angew Chem Int Ed 54:5451–5455
41. Fürstner A, Domostoj MM, Scheiper B (2005) J Am Chem Soc 127:11620–11621
42. Obradors C, Leboeuf D, Aydin J, Echavarren AM (2013) Org Lett 15:1576–1579

Chapter 3
Construction of the Pyrrolo[2,3-*d*]Carbazole Core of Spiroindoline Alkaloids by Gold-Catalyzed Cascade Cyclization of Ynamide

Abstract A novel gold-catalyzed cascade cyclization of ynamides for the construction of pyrrolo[2,3-*d*]carbazole scaffold was developed. This reaction proceeds through a formation of spiroindoline by 5-*exo* cyclization, followed by trapping of the resulting iminium intermediate. The cyclization was allowed for the synthesis of enantiomerically enriched pyrrolo[2,3-*d*]carbazole by using the chiral gold complex. This methodology provides an access to the asymmetric formal synthesis of vindorosine.

Vindorosine has a highly substituted spirocyclic indoline fused with a cyclohexane ring bearing continuous stereocenters (Fig. 3.1). The structural complexity and biological activities of vindorosine have inspired the synthetic community. As described in Chap. 1, various efficient total syntheses of vindorosine have been reported. However, to the best of our knowledge, the asymmetric syntheses reported to date have relied on chiral pool strategies or diastereoselective reactions with the use of chiral auxiliaries. Thus, the total synthesis of vindorosine based on a catalytic asymmetric reaction remains challenging. The author expected gold catalysis would provide an efficient approach to vindorosine and related alkaloids in an enantioselective manner.

Gold-catalyzed annulation of ynamides has emerged as a useful strategy for the construction of polycyclic nitrogen heterocycles [1–3]. Following Dankwardt's report, a number of effective synthetic approaches to the construction of cyclic structures based on a gold-catalyzed cyclization of alkyne and silyl enol ether have been developed [4]. The author's group previously demonstrated that the enynes containing silyl enol ether have highly potential as a building block for the functionalized indole synthesis (Scheme 3.1a) [5] In this reaction, 5-*endo-dig* hydroamination provided indole and subsequent 6-*exo-dig* carbocyclization from enol ether moiety proceeded to form piperidine ring in a one-pot manner. The author envisioned that a gold-catalyzed annulation of ynamides bearing a silyl enol ether moiety would provide a direct access to pyrrolo[2,3-*d*]carbazole, the common tetracyclic indoline core of aspidosperma and malagasy alkaloids. The author's working hypothesis is shown in Scheme 3.1b. Activation of the triple bond of ynamide **A** would promote nucleophilic attack at the indole 3 position to generate the spiroindoline intermediate **B**.

© The Editor(s) (if applicable) and The Author(s), under exclusive license to Springer Nature Singapore Pte Ltd. 2020
J. Matsuoka, *Total Synthesis of Indole Alkaloids*, Springer Theses, https://doi.org/10.1007/978-981-15-8652-1_3

67

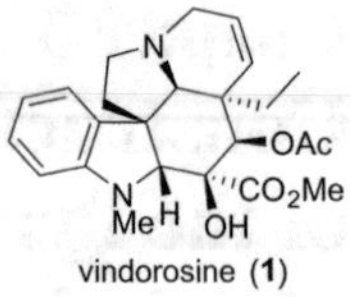

Fig. 3.1 Structure of
vindorosine

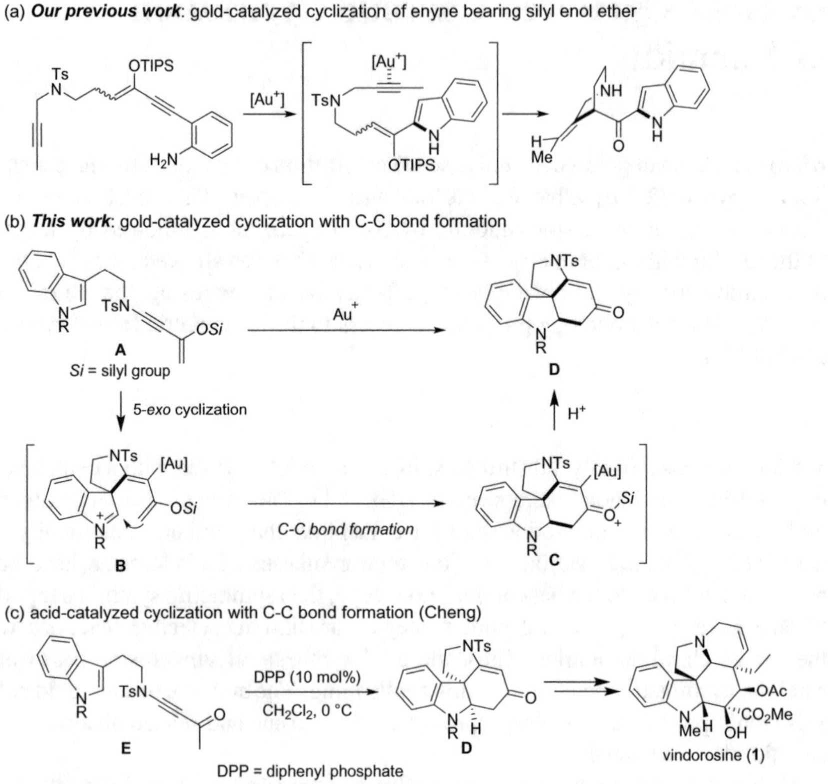

vindorosine (1)

(a) *Our previous work*: gold-catalyzed cyclization of enyne bearing silyl enol ether

(b) *This work*: gold-catalyzed cyclization with C-C bond formation

(c) acid-catalyzed cyclization with C-C bond formation (Cheng)

DPP = diphenyl phosphate

Scheme 3.1 Intramolecular cascade reaction of indole-ynamides

The subsequent addition of silyl enol ether to the resulting iminium moiety leads to the formation of intermediate **C**, followed by deauration and cleavage of the silyl group to produce pyrrolo[2,3-d]carbazole **D**. Quite recently, the Cheng and Liu group [6] reported an acid-catalyzed cascade reaction of ynamide for the racemic synthesis of pyrrolo[2,3-d]carbazole **D** using methyl ketone derivative **E** (Scheme 3.1c). In this chapter, the author describes a gold-catalyzed cascade reaction of ynamide with silyl enol ether, leading to pyrrolo[2,3-d]carbazole **D**. A catalytic enantioselective version of the reaction and its application to formal synthesis of **1** are also presented.

The preparation of silyl enol ether **7** is shown in Scheme 3.2. The protected tryptamine **3**, prepared by tosylation and benzylation of tryptamine **2**, was treated

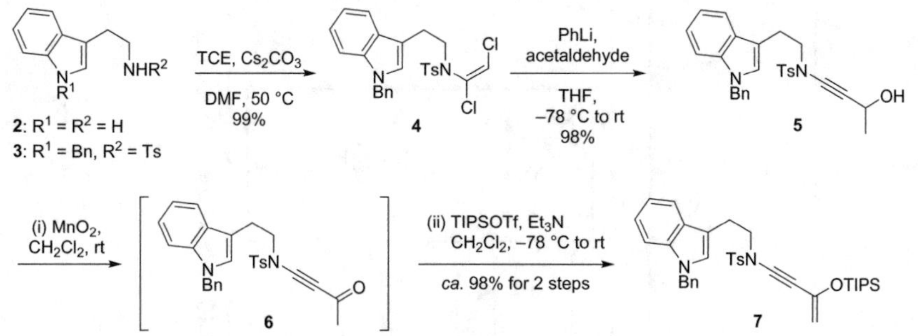

Scheme 3.2 Synthesis of indole-ynamide having a silyl enol ether

with trichloroethene (TCE) and Cs_2CO_3 to give dichloroenamine **4** in 99% yield. Dehydrochlorination–lithiation of **4** with PhLi and subsequent addition to acetalde-hyde resulted in ynamide **5** bearing a secondary hydroxy group in 98% yield [7]. Silyl enol ether **7** was then formed through a two-step sequence involving oxidation of **5** with MnO_2 and subsequent silylation of ketone **6**. Note that ketone **6** was not isolated because of its instability on silica gel [8].

The author then explored the optimal conditions for the gold-catalyzed cascade cyclization (Table 3.1). The treatment of methyl ketone **6** with JohnPhosAuSbF$_6$ led to recovery of the unreacted starting material without providing the desired product (entry 1). The author next performed the reaction of silyl enol ether **7** with JohnPhosAuSbF$_6$. Rewardingly, the expected cascade cyclization proceeded smoothly to afford the desired tetracyclic indoline **8** in 74% yield (entry 2). Consid-ering that the demetalation of the vinylgold intermediate of type **C** requires protona-tion [5], the author next evaluated the influence of additional proton sources. Among *i*-PrOH, AcOH, and TFA (entries 3–5), only *i*-PrOH had a positive effect on the yield of the desired product **8** (79%, entry 3). Finally, optimization of the ligands (entries 6–8) revealed that IPr was most effective in terms of the yield of the desired product (91%, entry 8).

Next the author proceeded to investigate the asymmetric gold-catalyzed cascade reaction. The author focused on the biaryl-type chiral ligands, which are known to act as efficient ligands for gold-catalyzed asymmetric reactions in the previous reports (Table 3.2) [3f, 9–11]. The author tested cationic binuclear gold complexes derived from chiral C_2-symmetrical bis-phosphine ligands **L2–5** in the cascade reaction (entries 1–4) and found that DTBM-BINAP (**L3**, entry 2) gave the most promising result (58% yield, 50% *ee*). Evaluation of the counterions using **L3** (entries 5–7) revealed that sodium tetrakis[3,5-bis(trifluoro-methyl)phenyl]borate (NaBARF) improved the enantioselectivity to 72% but decreased the yield of (−)-**8** to 38% (entry 7). A slight increase in enantioselectivity was observed when using a DTBM-SEGPHOS complex, **L5**Au$_2$Cl$_2$/NaBARF (38% yield, 74% *ee*, entry 8). Unfortu-nately, further investigations on the reaction temperature and catalyst loading did not further improve the yield.

Table 3.1 Optimization of the racemic reaction[a]

Entry	Substrate	LAuX[b]	AgY	Additive	Yield (%)[f]
1[c]	6[d]	L1AuSbF6	–	–	ND
2	7[e]	L1AuSbF6	–	–	74
3	7[e]	L1AuSbF6	–	i-PrOH	79
4	7[e]	L1AuSbF6	–	AcOH	61
5	7[e]	L1AuSbF6	–	TFA	68
6	7[e]	L1AuCl	AgSbF6	i-PrOH	63
7	7[e]	Ph3PAuCl	AgSbF6	i-PrOH	69
8	7[e]	IPrAuCl	AgSbF6	i-PrOH	91

[a]Reaction conditions: substrate (6 or 7, 1 equiv), Au(I)-ligand (5 mol%), AgY (5 mol%), 1,2-dichloroethane (DCE), additive (10 equiv where applicable), rt. [b]Catalysts were prepared in situ by mixing AuCl-ligand with AgY, except for JohnPhosAu(MeCN)SbF6 (prepared in advance). [c]Reaction was performed for 24 h. [d]Crude substrate was used owing to the instability of 6 on silica gel. [e]Including TIPSOH (9–15%). [f]Isolated yields based on the purity of 7 (85–91%)

Table 3.2 Optimization of the asymmetric reaction[a]

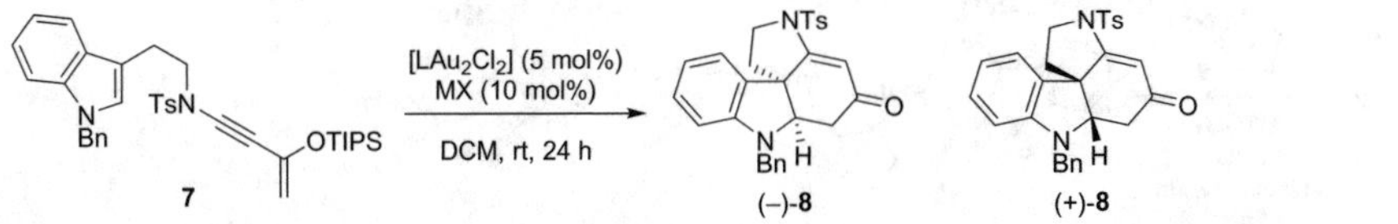
TsN
N
Bn
OTIPS
7
[LAu2Cl2] (5 mol%)
MX (10 mol%)
DCM, rt, 24 h
NTs
N
Bn H
O
(–)-8
(+)-8

Entry	$LAu_2Cl_2^b$	MX	Yield (%)[c]	% *ee*[d]
1	**L2**Au_2Cl_2	$AgBF_4$	trace	2 (–)
2	**L3**Au_2Cl_2	$AgBF_4$	58	50 (–)
3	**L4**Au_2Cl_2	$AgBF_4$	23	12 (–)
4	(*R*)-**L5**Au_2Cl_2	$AgBF_4$	30	52 (–)
5	**L3**Au_2Cl_2	$AgSbF_6$	32	54 (–)
6	**L3**Au_2Cl_2	$LiB(C_6F_5)_4$	17	38 (–)
7	**L3**Au_2Cl_2	NaBARF	38	72 (–)
8	(*R*)-**L5**Au_2Cl_2	NaBARF	38	74 (–)
9	(*S*)-**L5**Au_2Cl_2	NaBARF	33	72 (+)

[a]Reaction condition: **7** (including 9–15% TIPSOH, 1 equiv), Au(I)·ligand (5 mol%), MX (10 mol%), dichloromethane (DCM)
[b]Unless otherwise noted the catalysts were prepared in situ by mixing AuCl·ligand with MX. Ligand structures are shown below
[c]Isolated yields based on the purity of **7** (85–91%)
[d]Determined by chiral HPLC analysis

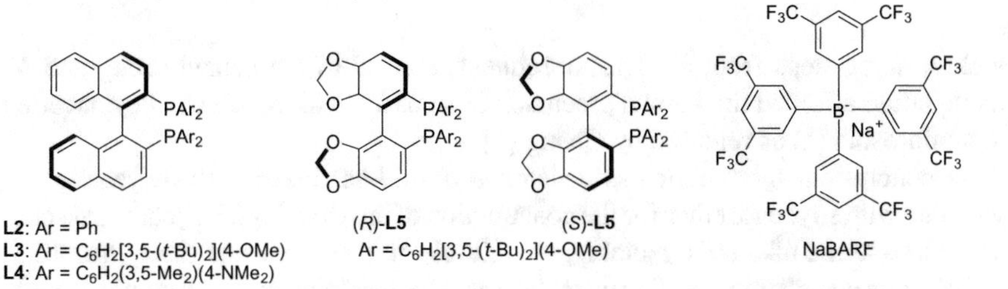
PAr2
PAr2
L2: Ar = Ph
L3: Ar = C6H2[3,5-(t-Bu)2](4-OMe)
L4: Ar = C6H2(3,5-Me2)(4-NMe2)
(R)-L5
(S)-L5
Ar = C6H2[3,5-(t-Bu)2](4-OMe)
F3C
CF3
B⁻
Na⁺
NaBARF

To reveal the absolute configuration of optically active **8** prepared by the reaction of **7** and a chiral gold catalyst, the author synthesized hydrazone **10** from racemic **8** and chiral hydrazine **9** (Scheme 3.3). After separation of diastereomers by column chromatography, the absolute configuration of **10** was confirmed by X-ray analysis. Then, we obtained the authentic sample of (*S*,*S*)–(+)-**8** by the reaction of **10** with MeI and H_2O. This experiment has revealed that (–)-**8** has the opposite configuration to that of natural vindorosine. Thus, we prepared (+)-**8** by the reaction using (*S*)-**L5**$AuCl_2$ (Table 3.2, entry 9).

The author finally applied the catalytic asymmetric synthesis of pyrrolo[2,3-*d*]carbazole **8** to formal total synthesis of vindorosine (Scheme 3.4). The annulation of **7** was carried out on a 1.3 mmol scale to produce (+)-**8** in 68% *ee* (21% isolated

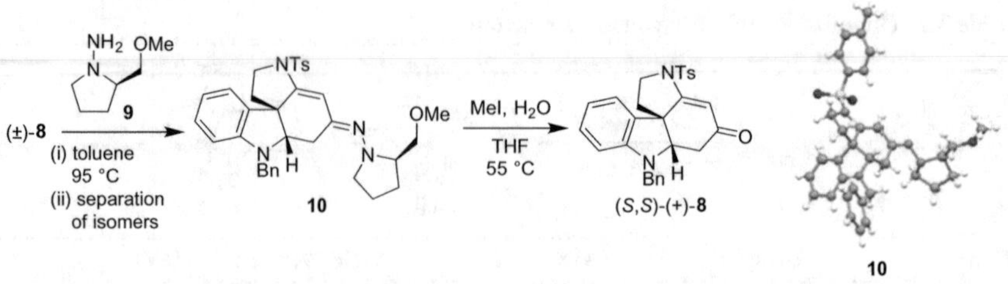

Scheme 3.3 Determination of absolute configuration of **8**

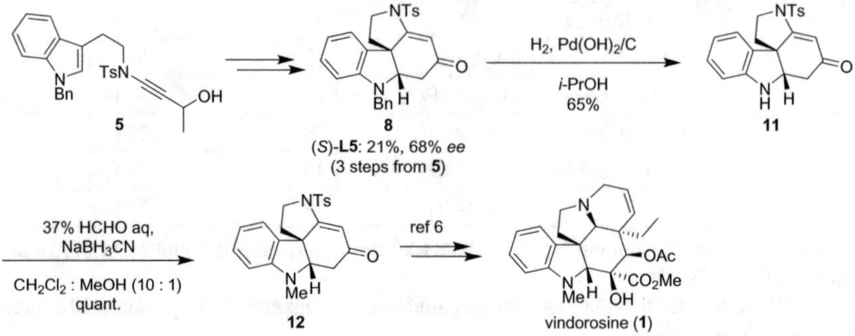

Scheme 3.4 Formal synthesis of vindorosine

yield in three steps from **5**). The subsequent removal of the benzyl group and *N*-methylation afforded the known precursor **12** in 65% yield, which can be converted to vindorosine (**1**) as reported by Cheng [6].

In conclusion, the author has developed a gold-catalyzed cascade reaction of ynamide with silyl enol ether for the construction of the pyrrolo[2,3-*d*]carbazole core of aspidosperma alkaloids. Notably, enantioselective synthesis of the pyrrolo[2,3-*d*]carbazole was also achieved in up to 74% *ee*. The developed reaction would provide access to spiroindoline alkaloids including vindorosine.

3.1 Experimental Section

3.1.1 General Methods

IR spectra were determined on a JASCO FT/IR-4100 spectrometer. Exact mass (HRMS) spectra were recorded on JMS-HX mass spectrometer or Shimadzu LC-ESI-IT-TOF-MS equipment. ^{1}H NMR spectra were recorded using a JEOL AL-400 or JEOL ECA-500. Chemical shifts are reported in δ (ppm) relative to Me$_4$Si (in CDCl$_3$) as an internal standard. ^{13}C NMR spectra were recorded using a

JEOL ECA-500 unit and referenced to the residual solvent signal. Melting points were measured by a hot stage melting point apparatus (uncorrected). For column chromatography, silica gel (Wakogel C-200E: Wako Pure Chemical Industries, Ltd) or amine silica gel (CHROMATOREX NH-DM1020: Fuji Silysia Chemical Ltd.) was employed. Chiral chromatography was performed with a Cosmosil CHiRAL 5B column (4.6 mm × 250 mm, Nacalai Tesque Inc.) or CHIRALCEL OD-H column (4.6 mm × 250 mm, Daicel Inc.) with using *n*-hexane/*i*-PrOH as an eluent. The gold complexes (*R*)-DTBM-SEGPHOS(AuCl)$_2$, (*S*)-DTBM-SEGPHOS(AuCl)$_2$, (*R*)-DTBM-BINOL(AuCl)$_2$, and (*R*)-DADMP-BINOL(AuCl)$_2$ were prepared according to the literature [10, 11].

3.1.2 Preparation of the Cyclization Precursor

N-[2-(1*H*-Indol-3-yl)ethyl]-4-methylbenzenesulfonamide (S1). To a solution of tryptamine (**2**) (9.61 g, 60.0 mmol) and Et$_3$N (9.20 mL, 66.0 mmol) in CH$_2$Cl$_2$ (90 mL) at 0 °C was added *p*-TsCl (13.7 g, 72.0 mmol) in one portion. After being stirred for 2 h at room temperature, the mixture was diluted with 1 M HCl and neutralized with 1 M NaOH. The resulting mixture was extracted with CH$_2$Cl$_2$ twice. The organic layer was washed with brine, dried over MgSO$_4$, filtered, and concentrated *in vacuo*. The residue was filtered through a short pad of NH$_2$ silica gel with CHCl$_3$ to afford **S1** (18.9 g, 60.0 mmol, 100%) as a white solid. This material was recrystallized from CHCl$_3$ to afford pure **S1** as colorless needles: mp 112–114 °C; IR (CDCl$_3$) 3406 (N–H), 1319 (O=S=O), 1152 (O=S=O); ^{1}H NMR (500 MHz, CDCl$_3$) δ 2.37 (s, 3H), 2.90 (t, *J* = 6.6 Hz, 2H), 3.24 (td, *J* = 6.6, 6.0 Hz, 2H), 4.58 (t, *J* = 6.0 Hz, 1H), 6.92 (d, *J* = 2.3 Hz, 1H), 7.04 (dd, *J* = 7.4, 7.4 Hz, 1H), 7.15–7.19 (m, 3H), 7.32 (d, *J* = 8.6 Hz, 1H), 7.39 (d, *J* = 7.4 Hz, 1H), 7.62 (d, *J* = 8.0 Hz, 2H), 8.12 (br s, 1H); ^{13}C NMR (125 MHz, CDCl$_3$) δ 21.4, 25.4, 43.0, 111.3, 111.4, 118.4, 119.4, 122.1, 122.6, 126.8, 127.0 (2C), 129.6 (2C), 136.3, 136.6, 143.3; HRMS (ESI) calcd for C$_{17}$H$_{18}$N$_2$NaO$_2$S$^+$ [M + Na]$^+$ 337.0981, found 337.0982.

N-[2-(1-Benzyl-1*H*-indol-3-yl)ethyl]-4-methylbenzenesulfonamide (3). To a solution of **S1** (18.9 g, 60.0 mmol) in dry DMF (200 mL) was slowly added NaH

(60% dispersion in mineral oil, 8.40 g, 210 mmol) at room temperature under argon, and stirring continued at this temperature for 30 min. The solution was cooled to 0 °C, and BnBr (7.09 mL, 60.0 mmol) was added dropwise. After being stirred for 2 h, the reaction mixture was allowed to warm up to room temperature and stirred overnight. The reaction mixture was diluted with sat. NH$_4$Cl and extracted with EtOAc. The combined organic layer was washed with water and brine and dried over MgSO$_4$. After concentration *in vacuo*, the residue was purified by flash chromatography on silica gel (hexane/EtOAc = 4/1 → 3/1) to afford **3** (19.2 g, 47.5 mmol, 79%) as a pale yellow solid: mp 90–93 °C; IR (CDCl$_3$) 3278 (N−H), 1323 (O =S=O), 1154 (O=S=O); ^{1}H NMR (500 MHz, CDCl$_3$) δ 2.38 (s, 3H), 2.91 (t, *J* = 6.9 Hz, 2H), 3.26 (td, *J* = 6.9, 5.7 Hz, 2H), 4.48 (d, *J* = 5.7 Hz, 1H), 5.23 (s, 2H), 6.85 (s, 1H), 7.04 (dd, *J* = 7.7, 7.7 Hz, 1H), 7.09 (d, *J* = 6.9 Hz, 2H), 7.14–7.19 (m, 3H), 7.24–7.31 (m, 4H), 7.41 (d, *J* = 8.0 Hz, 1H), 7.62 (d, *J* = 8.0 Hz, 2H); ^{13}C NMR (125 MHz, CDCl$_3$) δ 21.5, 25.4, 43.1, 49.9, 109.8, 110.7, 118.7, 119.2, 122.0, 126.5, 126.8 (2C), 127.0 (2C), 127.5, 127.7, 128.8 (2C), 129.6 (2C), 136.76, 136.78, 137.3, 143.2. *Anal.* calcd for C$_{24}$H$_{24}$N$_2$O$_2$S: C, 71.26; H, 5.98; N, 6.93. Found: C, 71.18; H, 6.00; N, 6.89.

(*E*)-*N*-[2-(1-Benzyl-1*H*-indol-3-yl)ethyl]-*N*-(1,2-dichlorovinyl)-4-methylbenzenesulfonamide (4). To a solution of **3** (19.2 g, 47.4 mmol) and Cs$_2$CO$_3$ (23.2 g, 52.1 mmol) in dry DMF (47 mL) was added dropwise trichloroethylene (4.69 mL, 52.1 mmol) over 10 min at room temperature under argon. The reaction mixture was allowed to warm up to 50 °C and stirred for 1.5 h. Upon cooling to room temperature, the reaction mixture was diluted with EtOAc, washed with water and brine, dried over MgSO$_4$, filtered, and concentrated *in vacuo*. The residue was filtered through a short pad of silica gel (EtOAc) to afford **4** (23.4 g, 46.9 mmol, 99%) as a pale yellow solid. This material was recrystallized from EtOAc to afford pure **4** as colorless needles: mp 103–107 °C; IR (CDCl$_3$) 1357 (O=S=O), 1164 (O=S=O); ^{1}H NMR (500 MHz, CDCl$_3$) δ 2.37 (s, 3H), 3.03 (t, *J* = 7.7 Hz, 2H), 3.53 (br s, 2H), 5.20 (s, 2H), 6.51 (s, 1H), 6.93 (s, 1H), 7.07–7.09 (m, 3H), 7.15 (dd, *J* = 7.7, 7.7 Hz, 3H), 7.18–7.28 (m, 4H), 7.51 (d, *J* = 8.0 Hz, 1H), 7.75 (d, *J* = 8.0 Hz, 2H); ^{13}C NMR (125 MHz, CDCl$_3$) δ 21.5, 24.0, 48.3, 49.8, 109.7, 110.6, 118.6, 119.2, 121.4, 121.8, 126.3, 126.8 (2C), 127.5, 127.7, 128.2 (2C), 128.7 (2C), 129.6 (3C), 135.0, 136.5, 137.3, 144.4. *Anal.* calcd for C$_{26}$H$_{24}$Cl$_2$N$_2$O$_2$S: C, 62.53; H, 4.84; N, 5.61. Found: C, 62.32; H, 4.77; N, 5.66.

N-[2-(1-Benzyl-1*H*-indol-3-yl)ethyl]-*N*-(3-hydroxybut-1-yn-1-yl)-4-methylbenzenesulfonamide (5). To a solution of **4** (2.50 g, 5.01 mmol) in dry THF (50 mL) was added PhLi (*ca.* 1.6 M in dibutyl ether, 6.88 mL, 11.0 mmol) dropwise at −78 °C under argon. The reaction mixture was stirred at this temperature for 2 h. After complete conversion to the intermediate (confirmed by TLC), acetaldehyde (0.340 mL, 6.10 mmol) was added at −78 °C and the reaction mixture was allowed to warm up to room temperature. After being stirred for 1 h, the reaction mixture was diluted with water and extracted with EtOAc. The combined organic layers were washed with water and brine and dried over MgSO$_4$. After concentration *in vacuo*, the residue was purified by flash chromatography on silica gel (hexane/EtOAc = 2/1) to afford **5** (2.31 g, 4.89 mmol, 98%) as an orange amorphous: IR (CDCl$_3$) 3407 (O–H), 2240 (C≡C), 1357 (O=S=O), 1165 (O=S=O); ^{1}H NMR (500 MHz, CDCl$_3$) δ 1.41 (d, J = 6.9 Hz, 3H), 2.00 (d, J = 5.2 Hz, 1H), 2.40 (s, 3H), 3.08 (t, J = 7.7 Hz, 2H), 3.57–3.66 (m, 2H), 4.57–4.62 (m, 1H), 5.21 (s, 2H), 6.89 (s, 1H), 7.09 (m, 3H), 7.16 (t, J = 7.7 Hz, 1H), 7.23–7.29 (m, 6H), 7.54 (d, J = 7.4 Hz, 1H), 7.71 (d, J = 8.0 Hz, 2H); ^{13}C NMR (125 MHz, CDCl$_3$) δ 21.6, 24.2, 24.3, 49.8, 51.7, 58.4, 73.0, 77.5, 109.7, 110.6, 118.7, 119.2, 121.8, 126.6, 126.8 (2C), 127.5 (2C), 127.6, 127.8, 128.7 (2C), 129.6 (2C), 134.6, 136.5, 137.7, 144.5; HRMS (ESI) calcd for C$_{28}$H$_{29}$N$_2$O$_3$S$^+$ [M + H]$^+$ 473.1893, found 473.1887.

N-[2-(1-Benzyl-1H-indol-3-yl)ethyl]-4-methyl-N-{3-[(triisopropylsilyl)oxy]but-3-en-1-yn-1-yl}benzenesulfonamide (7). To a solution of **5** (5.57 g, 11.8 mmol) in dry CH$_2$Cl$_2$ (118 mL) was added MgO (30.5 g, 353 mmol) at room temperature. After being stirred for 6 h, the reaction mixture was filtered through Celite. The filtrate was concentrated *in vacuo* to afford the corresponding ketone **6**. This material was used for the next reaction without further purification because of its instability toward silica gel. To a solution of the crude **6** and Et$_3$N (4.11 mL, 29.5 mmol) in dry CH$_2$Cl$_2$ (118 mL) was added TIPSOTf (3.96 mL, 14.7 mmol) dropwise at −78 °C under argon, and the reaction mixture was allowed to warm up to room temperature. After being stirred for 2 h, the reaction mixture was diluted with sat. NH$_4$Cl and extracted with CH$_2$Cl$_2$. The combined organic extracts were washed with water and brine and dried over Na$_2$SO$_4$. After concentration *in vacuo*, the residue was purified by flash chromatography on NH$_2$ silica gel (hexane/EtOAc = 12/1) to afford **7** (7.21 g, 11.5 mmol, *ca.* 98%; including a small amount of TIPSOH) as a pale yellow oil; IR (CDCl$_3$) 2229 (C≡C), 1369 (O=S=O), 1167 (O=S=O); ^{1}H NMR (500 MHz, CDCl$_3$) δ 1.09 (d, J = 7.5 Hz, 18H), 1.18–1.26 (m, 3H), 2.41 (s, 3H), 3.10 (t, J = 8.1 Hz, 2H), 3.64 (t, J = 7.8 Hz, 2H), 4.63 (s, 1H), 4.72 (s, 1H), 5.24 (s, 2H), 6.92 (s, 1H), 7.08–7.12 (m, 3H), 7.17 (t, J = 7.0 Hz, 1H), 7.23–7.31 (m, 6H), 7.55 (d, J = 7.5 Hz, 1H), 7.71 (d, J = 8.1 Hz, 2H); ^{13}C NMR (125 MHz, CDCl$_3$) δ 12.5 (3C), 17.9 (6C), 21.6, 24.3, 49.9, 52.0, 69.1, 79.8, 102.6, 109.7, 110.6, 118.7, 119.2, 121.9, 126.4, 126.8 (2C), 127.5 (2C), 127.6, 127.8, 128.7 (2C), 129.7 (2C), 134.7, 136.5, 137.4, 139.6, 144.5; HRMS (ESI) calcd for C$_{37}$H$_{47}$N$_2$O$_3$SSi$^+$ [M + H]$^+$ 627.3071, found 627.3071.

3.1.3   Gold-Catalyzed Cascade Cyclization

(6aR*,11bR*)-7-Benzyl-3-tosyl-2,3,6a,7-tetrahydro-1H-pyrrolo[2,3-d]carbazol-5(6H)-one (8) (Table 3.1, entry 8). To a solution of **7** (63 mg, 0.10 mmol) in DCE (1.0 mL) was added IPrAuCl (3.1 mg, 5.0 μmol), AgSbF$_6$ (1.7 mg, 5.0 μmol), and *i*-PrOH (77 μL, 1.0 mmol) at room temperature. After the mixture was stirred for 6 h, TBAF (*ca.* 1 M in THF, 0.15 mL, 0.15 mmol) was added. After being stirred for 30 min at room temperature, the reaction mixture was concentrated *in vacuo*. The residue was purified by flash chromatography on silica gel (hexane/EtOAc = 3/1) to afford (+)-**8** (43 mg, 0.091 mmol, 91%) as a yellow solid: mp 180–182 °C; IR (CDCl$_3$) 1620 (C=O), 1360 (O=S=O), 1168 (O=S=O);

^{1}H NMR (500 MHz, CDCl$_3$) δ 1.98 (ddd, J = 11.7, 11.7, 8.2 Hz, 1H), 2.10–2.15 (m, 2H), 2.49 (s, 3H), 2.51 (dd, J = 16.6, 6.3 Hz, 1H), 3.76 (dd, J = 10.3, 5.7 Hz, 1H), 3.80 (td, J = 10.9, 5.3 Hz, 1H), 3.96 (d, J = 14.9 Hz, 1H), 4.04 (dd, J = 10.0, 8.3 Hz, 1H), 4.43 (d, J = 14.9 Hz, 1H), 6.02 (d, J = 6.9 Hz, 1H), 6.29 (s, 1H), 6.46 (m, 2H), 7.06–7.10 (m, 1H), 7.27–7.30 (m, 1H), 7.32–7.37 (m, 4H), 7.41 (d, J = 8.0 Hz, 2H), 7.88 (d, J = 8.6 Hz, 2H); ^{13}C NMR (125 MHz, CDCl$_3$) δ 21.6, 35.2, 35.4, 48.7, 48.9, 53.8, 67.4, 106.5, 109.1, 118.7, 122.0, 127.2 (2C), 127.6, 127.8 (2C), 128.7 (2C), 129.1, 130.3 (2C), 130.5, 134.6, 136.9, 145.5, 148.1, 159.4, 196.4; HRMS (ESI) calcd for C$_{28}$H$_{27}$N$_2$O$_3$S$^+$ [M + H]$^+$ 471.1737, found 471.1738.

Asymmetric reaction using DTBM-SEGPHOS(AuCl)$_2$ (Table 3.2, entry 9)

To a solution of **7** (63 mg, 0.10 mmol) in DCM (1.0 mL) was added (*S*)-DTBM-SEGPHOS(AuCl)$_2$ (8.2 mg, 5.0 μmol) and NaBARF (8.9 mg, 1.0 μmol) at room temperature. After being stirred for 24 h, the reaction mixture was concentrated *in vacuo*. The residue was purified by flash chromatography on silica gel (hexane/EtOAc = 3/1) to afford **8** (18 mg, 0.038 mmol 33%, 72% *ee*) as a yellow amorphous solid [HPLC, CHIRALCEL OD-H column eluting with 65% *i*-PrOH/*n*-hexane over 30 min at 0.80 mL/min, t_1 = 17.66 min (minor isomer), t_2 = 25.44 min (major isomer)].

(6a*S*,11b*S*,*E*)-7-Benzyl-*N*-[(*R*)-2-(methoxymethyl)pyrrolidin-1-yl]-3-tosyl-2,3,6a,7-tetrahydro-1*H*-pyrrolo[2,3-*d*]carbazol-5(6*H*)-imine (10). A mixture of racemic **8** (0.10 g, 0.21 mmol) and (*R*)-1-amino-2-(methoxymethyl)pyrrolidine **9** (56 μL, 0.043 mmol) in toluene (1.0 mL) was stirred at 95 °C for 24 h. The reaction mixture was concentrated *in vacuo*. The residue was purified by flash chromatography on amine silica gel (hexane/EtOAc = 4/1) to afford **10** (9.6 mg, 0.016 mmol, 8%) as a white solid. This material was recrystallized from MeCN: mp 88–92 °C; [α]$_D^{26}$ −206.2 (c 0.48, CHCl$_3$); IR (CDCl$_3$) 1643, 1600 (C=N), 1355 (O=S=O), 1166 (O=S=O); ^{1}H NMR (500 MHz, CDCl$_3$) δ 1.64–1.73 (m, 2H), 1.78–1.84 (m, 2H), 1.90–2.06 (m, 3H), 2.29–2.34 (m, 1H), 2.48 (s, 3H), 3.07–3.11 (m, 1H), 3.20–3.27 (m, 2H), 3.33–3.36 (m, 4H), 3.41–3.43 (dd, J = 9.0, 4.0 Hz, 1H), 3.57–3.65 (m, 2H), 3.95 (t, J = 8.0 Hz, 1H), 4.23 (d, J = 15.0 Hz, 1H), 4.36 (d, J = 15.0 Hz, 1H), 5.65 (d, J = 7.0 Hz, 1H), 6.31–6.35 (m, 2H), 6.54 (s, 1H), 6.98 (t, J = 8.0 Hz, 1H), 7.27–7.37 (m, 5H), 7.40 (d, J = 8.0 Hz, 2H), 7.90 (d, J = 8.0 Hz, 2H); ^{13}C NMR (125 MHz, CDCl$_3$) δ 21.6, 22.5, 25.7, 26.6, 36.1, 47.7, 49.2, 53.3, 54.4, 59.2, 66.6, 67.8, 75.4, 107.8, 108.7, 117.9, 122.2, 127.27 (2C), 127.30 (3C), 128.4, 128.6 (2C), 130.0 (2C), 132.4, 135.3, 137.9, 144.3, 144.5, 148.2, 158.3; HRMS (ESI) calcd for C$_{34}$H$_{39}$N$_4$O$_3$S$^+$ [M + H]$^+$ 583.2737, found 583.2735.

(6a*S*,11b*S*)-7-Benzyl-3-tosyl-2,3,6a,7-tetrahydro-1*H*-pyrrolo[2,3-*d*]carbazol-5(6*H*)-one [(*S*,*S*)-(+)-8]. A mixture of **10** (5.8 mg, 0.010 mmol) and MeI (6.2 μL, 0.10 mmol) in THF (0.20 mL) was stirred at 55 °C for 48 h. The reaction mixture was concentrated *in vacuo*. The reaction mixture was diluted with water and extracted with EtOAc. The combined organic layers were washed with brine and dried over Na$_2$SO$_4$. After concentration *in vacuo*, the residue was purified by flash chromatography on silica gel (hexane/EtOAc = 2/1) to afford **8** (2.0 mg, 4.3 μmol, 43%): [α]$_D^{26}$ +84.3 (c 0.13, CHCl$_3$).

3.1.4 Formal Synthesis of Vindorosine

(6aS,11bS)-3-Tosyl-2,3,6a,7-tetrahydro-1H-pyrrolo[2,3-d]carbazol-5(6H)-one (11). A mixture of **8** (91 mg, 0.19 mmol) and Pd(OH)$_2$/C (ca. 50 wt% on carbon, 53 mg) in *i*-PrOH (1.0 mL) was stirred under a hydrogen atmosphere at 55 °C for 12 h. The resulting suspension was filtered through a Celite pad, and the pad was washed with EtOAc. The filtrate was concentrated *in vacuo*. The residue was purified by flash chromatography on silica gel (CHCl$_3$/MeOH = 20/1) to afford **11** (21 mg, 0.055 mmol, 29%, 65% 2 cycles) as a white solid: mp 232–237 °C; IR (CDCl$_3$) 1612 (C=O), 1360 (O=S=O), 1166 (O=S=O); ^{1}H NMR (500 MHz, CDCl$_3$) δ 1.98–2.07 (m, 2H), 2.23 (dd, J = 16.0, 10.0 Hz, 1H), 2.50 (s, 3H), 2.54 (dd, J = 16.5, 6.5 Hz, 1H), 3.76–3.82 (m, 2H), 3.93 (dd, J = 9.5, 6.5 Hz, 1H), 4.03–4.06 (m, 1H), 6.02 (d, J = 7.5 Hz, 1H), 6.33 (s, 1H), 6.49–6.52 (m, 1H), 6.72 (d, J = 8.0 Hz, 1H), 7.07 (ddd, J = 7.5, 7.5, 1.5 Hz, 1H), 7.43 (d, J = 8.0 Hz, 2H), 7.90 (d, J = 8.0 Hz, 2H); ^{13}C NMR (125 MHz, CDCl$_3$) δ 21.7, 35.4, 40.8, 48.6, 54.8, 63.8, 106.5, 111.7, 119.7, 122.3, 127.3 (2C), 129.1, 129.8, 130.3 (2C), 134.7, 145.5, 147.5, 159.2, 196.2; HRMS (ESI) calcd for C$_{21}$H$_{21}$N$_2$O$_3$S$^+$ [M + H]$^+$ 381.1267, found 381.1266.

(6aS,11bS)-7-Methyl-3-tosyl-2,3,6a,7-tetrahydro-1H-pyrrolo[2,3-d]carbazol-5(6H)-one (12). A mixture of **11** (21 mg, 0.055 mmol) and 37% HCHO aq (0.17 mL) in CH$_2$Cl$_2$: MeOH (10: 1) was added to NaBH$_3$CN (14 mg, 0.22 mmol) at 0 °C. The reaction mixture was adjusted to pH 3 and stirred for 1.5 h. The reaction mixture was diluted with NaHCO$_3$ and extracted with CH$_2$Cl$_2$. The combined organic extracts were washed with brine and dried over Na$_2$SO$_4$. After concentration *in vacuo*, the residue was purified by flash chromatography on silica gel (CHCl$_3$/MeOH = 40/1) to afford **12** (22 mg, 0.055 mmol, quant., 74% ee) as a white solid: [HPLC, Cosmosil CHiRAL 5B column eluting with 55% *i*-PrOH/*n*-hexane over 30 min at 0.80 mL/min, t_1 = 17.84 min (minor isomer), t_2 = 19.18 min (major isomer)]: mp 174–177 °C; IR (CDCl$_3$) 1616 (C=O), 1357 (O=S=O), 1168 (O=S=O); ^{1}H NMR (500 MHz, DMSO-d_6) δ 1.75 (dd, J = 17.0, 9.5 Hz, 1H), 1.86 (dd, J = 12.0, 5.0 Hz, 1H), 2.10–2.28 (m, 1H), 2.45–2.48 (m, 4H), 2.67 (s, 3H), 3.70–3.76 (m, 1H), 4.01 (dd, J = 10.0, 6.0 Hz, 1H), 4.07 (dd, J = 10.0, 8.0 Hz, 1H), 5.85 (d, J = 7.5 Hz, 1H), 5.93 (s, 1H), 6.40 (ddd, J = 7.0, 7.0, 1.0 Hz, 1H), 6.56 (d, J = 8.0 Hz, 1H), 7.08 (ddd, J = 7.5, 7.5, 1.0 Hz, 1H), 7.57 (d, J = 8.0 Hz, 2H), 7.96 (d, J = 8.5 Hz, 2H); ^{13}C NMR (125 MHz, DMSO-d_6) δ 21.1, 31.4, 34.59, 34.62, 48.8, 53.4, 68.3, 105.5, 108.7, 117.9, 121.2, 127.2 (2C), 129.0, 130.6 (3C), 134.1, 145.8, 149.0, 159.4, 195.2; HRMS (ESI) calcd for C$_{22}$H$_{23}$N$_2$O$_3$S$^+$ [M + H]$^+$ 395.1424, found 395.1425. The spectral data were in good agreement with those previously reported.[6]

3.1.5 Crystallography

The data of the compound **10** ($C_{34}H_{38}N_4O_3S$) were collected with a Rigaku XtaLAB P200 diffractometer using multilayer mirror monochromated Cu–Kα radiation at 93 K. The substance was crystallized from MeCN as clear block crystals and solved in primitive orthorhombic space group $P12_1/c1$ with $Z = 4$. The unit cell dimensions are $a = 9.1245(1)$, $b = 23.3744(1)$, $c = 14.4458(1)$, $V = 3002.41(4)$ Å^3, Dcalc $= 1.289$ g/cm^3, Mw: 582.74. $R = 0.0398$, GOF $= 1.049$, Flack parameter $= 0.004(7)$. The CCDC deposition number: CCDC 1911307.

References

1. For a review article on ynamides, see: **a** DeKorver KA, Li H, Lohse AG, Hayashi R, Lu Z, Zhang Y, Hsung RP (2016) Chem Rev 110:5064–5106; **b** Pan F, Shu C, Ye L-W (2016) Org Biomol Chem 14:9456–9465
2. For gold-catalyzed reaction of ynamides with azides, see **a** Shu C, Wang Y-H, Zhou B, Li X-L, Ping Y-F, Lu X, Ye L-W J Am Chem Soc 137:9567–9570; **b** Pan Y, Chen G-W, Shen C.-H, He W, Ye LW (2017) Org Chem Front 3:491–495; **c** Shen W-B.; Sun Q, Li L, Liu X, Zhou B, Yan J-Z, Lu X, Ye L-W (2017) Nat Commun 8:1748–1757
3. For recent reports on related gold-catalyzed cyclization of ynamides, see **a** Pan F, Liu S, Shu C, Lin R-K, Yu Y-F, Zhou J-M, Ye L-W (2014) Chem Commun 50:10726–10729; **b** Adcock HV, Chatzopoulou E, Davies PW (2015) Angew Chem Int Ed 54:15525–15529; **c** Liu J, Chen M, Zhang L, Liu Y (2015) Chem Eur J 21:1009–1013, **d** Nayak S, Ghosh N, Prabagar B, Sahoo AK (2015) Org Lett 17:5662–5665. For a related gold-catalyzed oxidative reaction, see **e** Lin M, Zhu L, Xia J, Yu Y, Chen J, Mao Z, Huang (2018) Adv Synth Catal 360:2280–2284; **f** Zheng N, Chang Y-Y, Zhang L-J, Gong J-X, Yang Z (2016) Chem Asian J 11:371–375
4. **a** Dankwardt JW (2001) Tetrahedron Lett 42:5809–5812; **b** Nevado C, Cárdenas DJ, Echavarren AM Chem Eur J 9:2627–2635; **c** Staben ST, Kennedy-Smith JJ, Huang D, Corkey BK, LaLonde RL, Toste FD (2006) Angew Chem Int Ed 45:5991–5994; **d** Minnihan EC, Colletti SL, Toste FD, Shen HC (2007) J Org Chem 72:6287–6289; **e** Ito H, Ohmiya H, Sawamura M (2010) Org Lett 12:4380–4383
5. Naoe S, Yoshida Y, Oishi S, Fujii N, Ohno H (2016) J Org Chem 81:5690–5698
6. Wang Y, Lin J, Wang X, Wang G, Zhang X, Yao B, Zhao Y, Yu P, Lin B, Liu Y, Cheng M (2018) Chem Eur J 24:4026–4032
7. Mansfield SJ, Campbell CD, Jones MW, Anderson EA (2015) Chem Commun 51:3316–3319

8. Preparation of the corresponding TBS-protected enol ether failed because of its lower stability compared with the TIPS derivative **7**. Similarly, preparation of the *N*-methyl indole derivative was difficult due to the instability of the *N*-methyl derivative of type **5** under oxidation conditions

9. For a review article, see **a** Wang Y, Lackner AD, Toste FD (2014) Acc Chem Res 47:889–901. For selected gold-catalyzed asymmetric reactions, see **b** LaLonde RL, Sherry BD, Kang EJ, Toste FD (2007) J Am Chem Soc 129:2452–2453; **c** Bandini M, Eichholzer A (2009) Angew Chem Int Ed 48:9533–9537; **d** Sethofer SG, Mayer T, Toste FD (2010) J Am Chem Soc 132:8276–8277; **e** González AZ, Toste FD (2010) Org Lett 12:200–203; **f** Brazeau J, Zhang S, Colomer I, Corkey BK, Toste FD (2012) J Am Chem Soc 134:2742–2749; **g** Cera G, Chiarucci M, Mazzanti A, Mancinelli M, Bandini M, Org Lett 14:1350–1353; **h** Wang Z, Nicolini C, Hervieu C, Wong Y, Zanoni G, Zhang L (2017) J Am Chem Soc 139:16064–16067

10. Johansson MJ, Gorin DJ, Staben ST, Toste FD (2005) J Am Chem Soc 127:18002–18003

11. Melhado AD, Luparia M, Toste FD (2007) J Am Chem Soc 129:12638–12639

Chapter 4
Conclusion

1. The total and formal syntheses of dictyodendrins A–F have been achieved. The key step in this process involved the direct construction of the pyrrolo[2,3-c]carbazole core by the gold-catalyzed annulation of a conjugated diyne with a pyrrole to form three bonds and two aromatic rings. The subsequent introduction substituents at the C1 (Suzuki–Miyaura coupling), C2 (acylation), N3 (alkylation), and C5 (Ullman coupling) positions provided divergent access to dictyodendrins. Comparing the reported syntheses of dictyodendrins, the author's synthetic procedure, based on construction of the pyrrolo[2,3-c]carbazole scaffolds followed by the introduction of substituents (C1, C2, N3, C5 position), allowed for the introduction of the various substituents in the late stage in the synthesis, thus providing divergent access to natural dictyodendrins and unnatural analogues with minimal efforts. Additionally, the assessment of biological activities revealed that dictyodendrin analogues were found to be a potential inhibitor of CDK2/CycA2 and GSK3.

2. Direct construction of the common pyrrolo[2,3-d]carbazole core of related alkaloids by a gold-catalyzed cascade cyclization of ynamide was developed. This reaction involves intramolecular cyclization from indole to ynamide followed by trapping of the resulting iminium intermediate. Through the use of chiral gold complexes, an enantiomerically enriched pyrrolo[2,3-d]carbazole was obtained in up to 74% ee. This methodology was successfully applied to the asymmetric formal synthesis of vindorosine. In the reported asymmetric syntheses of vindorosine, a chiral pool and chiral auxiliaries have been used for a chiral source. The author's strategy is the first example of asymmetric synthesis of vindorosine using a chiral catalyst as a chiral source.

In summary, the author developed two novel strategies for construction of pyrrolo-carbazole core structures on the basis of gold-catalyzed reactions of diynes or ynamides for efficient total synthesis of dictyodendrins and aspidosperma alkaloids. These finding would provide the access to the divergent-oriented synthetic strategy

J. Matsuoka, *Total Synthesis of Indole Alkaloids*, Springer Theses, https://doi.org/10.1007/978-981-15-8652-1_4

for the syntheses of related pyrrolecarbazole compounds. These methodologies accelarate the drug discovery study on the basis of alkaloid compounds.